Synthesis Lectures on Ocean Systems Engineering

Series Editor

Nikolas Xiros, University of New Orleans, New Orleans, LA, USA

The series publishes short books on state-of-the-art research and applications in related and interdependent areas of design, construction, maintenance and operation of marine vessels and structures as well as ocean and oceanic engineering.

Huidong Zhang · Carlos Guedes Soares

Numerical Modelling of Extreme Waves

Springer

Huidong Zhang
Ocean University of China
Qingdao, China

Carlos Guedes Soares
Universidade de Lisboa
Lisbon, Portugal

ISSN 2692-4420 ISSN 2692-4471 (electronic)
Synthesis Lectures on Ocean Systems Engineering
ISBN 978-3-031-77086-9 ISBN 978-3-031-77084-5 (eBook)
https://doi.org/10.1007/978-3-031-77084-5

This Springer imprint is published by the registered company Springer Nature Switzerland AG
The registered company address is: Gewerbestrasse 11, 6330 Cham, Switzerland

Preface

With more frequent extreme weather due to global warming and more rapid exploration and exploitation of the ocean by human beings, the chance of encountering an extreme wave is significant today. The destructive power of extreme waves is genuinely remarkable, and sometimes, they can cause damage to floating structures such as ships in the ocean, leading to severe loss of life and property or even to local pollution and environmental risk. Therefore, there is an urgent need in the industry community to find numerical models that can be used to predict the formation of extreme waves efficiently with acceptable accuracy.

The primary motivation for writing this book is to systematically exploit the capability of a series of numerical wave models with different order of nonlinearities in simulating extreme waves that appear in more realistic sea states by comparing numerical simulations with laboratory observations.

Moreover, to indicate the discrepancies in numerical simulations performed with various wave models, the corresponding numerical results for the same sea state are analysed statistically.

In addition, the influence of nonlinearities of different order on the generation of extreme waves is explicitly revealed in this book, and the exceedance probability of extreme wave height has been compared with various theoretical prediction models.

Besides, the effect of the synthesized initial conditions on the formation of extreme waves is comprehensively discussed from two aspects: the disturbance of noise in a deterministic wave sequence and the influence of three types of uncertainty in a stochastic wave train.

Qingdao, China — Huidong Zhang
Lisbon, Portugal — Carlos Guedes Soares

Acknowledgements

We want to thank many people for contributing to this book in different ways. We are very grateful to Prof. Miguel Onorato in Italy, who was the first coauthor of our wave modelling research. He provided valuable codes for our initial study on the NLS equation. He guided us in the following research of the Dysthe model and granted us access to the precious data from 3D laboratory experiments obtained in the joint project with Prof. Alessandro Toffoli in Australia.

Professor Alexander Babanin in Australia gave the first author an exchange study opportunity at Swinburne University of Technology. He introduced us to Prof. Dmitry Chalikov in Russia, who shared his method and helped us set up the fully nonlinear CS model.

We also appreciate the detailed explanation and valuable suggestion on numerical instabilities from Dr. Odin Gramstad in Norway and Professor Lev Shemer in Israel, and we are indebted to Prof. Guillaume Ducrozet in France for the help in using his open-source HOS method and Prof. Amin Chabchoub in Australia for the fruitful instruction on Peregrine breather solution. We have to thank Prof. Alexey Slunyaev in Russia for his help in solving several complicated mathematical equations.

Regarding the wave statistics, we thank Dr. Elzbieta Bitner-Gregersen in Norway, Dr. Zhivelina Cherneva, and Dr. Petya Petrova for their discussions in Portugal. We also appreciate Profs. Günther Clauss and Dr. Marco Klein in Germany for sharing the experimental data related to breather solutions.

The financial supports from the CENTEC research centre at the University of Lisbon, the Portuguese Foundation for Science and Technology (FCT) in Portugal, the National Natural Science Foundation of China, and the Ocean University of China are appreciated.

A special thanks must go to the first author's parents and friends for their permanent mental support and understanding of the decision to finish the challenging academic endeavour.

Contents

1 Introduction

1.1 Motivation

1.1.1 Threat of Extreme Waves

With the quick development of modern ocean projects and the more frequent appearance of extreme weather, scientists and engineers have begun to pay a lot of attention to the safety and reliability of marine and offshore structures because casualties have happened too frequently in the past decades.

It has been reported that many vessels sunk or were seriously damaged during the period from 1969 to 1994, mainly due to encountering the sudden occurrence of very large waves (see Fig. 1.1). Consequently, many people perished, cargoes were lost (see Fig. 1.2), and sometimes there was also severe pollution to the local ocean environment (Kharif et al. 2009).

In addition, offshore structures are more vulnerable to very large waves because they are not easy to move from one place to another. For example, on 15 February 1982, a giant wave suddenly smashed the windows and flooded the control room in a drilling rig operated by Mobil Oil on the Grand Banks of Newfoundland. Shortly afterwards, the rig capsized and sank, killing all 84 people on board (Lawton 2001).

Another widely known example is the Draupner Jacket platform, which was hit by the famous New Year Wave presented in Fig. 1.3 on 1 January 1995 with a height close to 26 m. However, the typical surrounding waves were only about 11–12 m and the estimated maximum expected wave height was supposed to be about 20 m.

In most recent literature, abnormal, rogue or freak waves are defined as those extraordinarily large water waves whose heights exceed two times the significant wave height in a sea state. This criterion has gained wide acceptance despite its shortcomings in the

H. Zhang and C. Guedes Soares, *Numerical Modelling of Extreme Waves*, Synthesis Lectures on Ocean Systems Engineering, https://doi.org/10.1007/978-3-031-77084-5_1

Fig. 1.1 Rogue waves taken on board. Left, taken by Philippe Lijour on the oil tanker Esso Languedoc in 1980; right, taken by a crewmember on the SS Spray in 1986

lack of additional criteria, such as the restriction on crest height (see, e.g., Clauss 2002; Kharif et al. 2009; Guedes Soares et al. 2011).

In most cases, giant water waves are generated in storms and hurricanes, meaning that rogue waves are much more probable under extreme weather conditions (Guedes Soares et al. 2004). The term "extreme waves" is used to specify the tail of some typical statistical distribution of wave heights, meaning that abnormal waves appear rarely but more often than expected from the traditional probability theory.

In this sense, relatively high waves are expected to be recorded in most marine incidents. However, Toffoli et al. (2005) found that relatively low significant wave heights were detected during some ship accidents that were reported as being due to bad weather. Hence, this suggests that some abnormal waves may have occurred, causing damage even in low sea states.

Indeed, the wave impact upon marine structures may also be determined by other parameters, such as wave length, wave steepness, crest height or horizontal wave asymmetry. Different types of marine structures may suffer from different wave parameters and conditions. For example, fishing vessels mainly capsize while fishing or loading fish (Toffoli et al. 2005). Until now, research on the dynamics of extreme waves and their effects on ships or platforms is still the primary interest of scientists and engineers.

Apart from the traditional tool of wave rider buoys, observations and measurements of extreme waves from space have become possible. However, the detailed record of extreme

Fig. 1.2 Examples of damage caused by the extreme waves (from Olagnon 2000)

Fig. 1.3 Time history recorded at Draupner platform on 01/01/1995 15:20:09 (from Taylor et al. 2006)

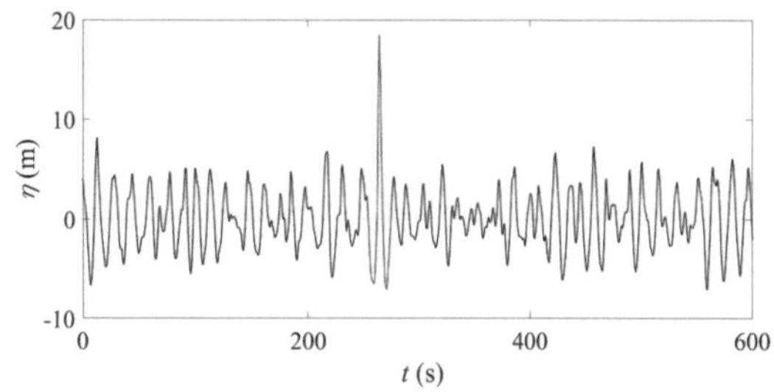

waves is minimal due to the rare probability of observing them and the harsh conditions to measure them in storms (Guedes Soares et al. 2011).

Alternatively, model tests can represent an essential source in studying their physical and statistical properties. However, laboratory experiment costs are usually very high, and the wave basins are not generally available to study only waves. Hence, there is a substantial need for reliable and computationally efficient numerical tools in water wave research.

Moreover, to test the performance of marine structures in the presence of extreme waves, numerous attempts have been made to explore the possibility of using deterministic nonlinear wave theories to predict the evolution of a random wave field and the generation of extreme waves.

Finally, it is necessary to predict the likelihood of encountering such waves and their induced effects on the structure to develop a consistent approach for designing marine structures under the loads resulting from extreme waves. One could anticipate that if these questions were solved, there might be a set of designed wave series that would represent a collection of possible extreme waves, and the relative performance of different structural designs could be compared based on their responses to these wave series.

In a word, the rapid developments in understanding physics and simulating the generation of extreme waves are not only due to scientific interest but also the strong requirement for practical application in the design and safe operation of ships and offshore structures.

1.1.2 Physical Mechanisms of Extreme Wave Formation

Although the extreme waves had been part of marine folklore for centuries, oceanographers did not believe them until the 70s of the last century. Intensive research has been motivated by the published record of the New Year Wave. During the previous 40 years, much progress has been made in understanding the physical mechanisms of extreme waves.

Under many circumstances, rogue or abnormal waves can be generated, and the mechanisms involved are various, which may be traced back to the following three types: geometrical focusing, dispersive focusing and nonlinear focusing (Dysthe et al. 2008).

Some reviews on the physical mechanisms of the extreme wave phenomenon have been made in the past. From a popular and physically general point of view, Slunyaev et al. (2011) described the extreme waves that appeared in deepwater, shallow-water and coastal areas, emphasizing the difference between water wave physics at these locations. To understand these complex linear and nonlinear mechanisms, one can seek the professional and comprehensive review of Kharif and Pelinovsky (2003).

Rogue waves are ubiquitous and have been observed in various contexts. The associated review has been done by Onorato et al. (2013a), who conducted a detailed discussion on rogue waves appearing in water and optics. The universal mechanism that may be used to explain these rogue waves is modulational instability. After reviewing some early instructive works in different physical situations, developed independently in the 1960s and early 1970s, Zakharov and Ostrovsky (2009) pointed out that various methods had been used to treat modulational instability, which might be formally different but essentially led to the same mathematical and physical results.

Restricting the formation of extreme waves to surface waves in the open sea, the mechanism of the formation of extreme waves is more complex because wind forcing, wave breaking, wave-current interactions, and the tridimensionality of wave fields cannot be neglected. As summarized by Dysthe et al. (2008), there is no generally accepted explanation or theory for the occurrence of such rogue waves.

However, some attempts to forecast extreme waves in a deterministic or a statistical fashion have been made, such as the warning system developed by ECMWF based on wave spectral evolution (Janssen and Bidlot 2009). Recently, some new perspectives on the formation of extreme waves were proposed by Onorato and Suret (2016), who believed that external factors could set the spectrum out of local equilibrium and that the induced nonlinearity in some cases could lead to substantial deviations from Gaussian statistics, and eventually to rogue waves.

Nevertheless, modulational instability or Benjamin-Feir instability (BFI) has received much attention. It is the only nonlinear mechanism that can catch the main features of rogue waves in the deep sea (Kharif et al. 2009).

Since the landmark discovery of modulational instability of weakly nonlinear deepwater wave trains by Lighthill and Benjamin & Feir, research associated with nonlinear deepwater waves has entered a new era. The nonlinear stage of the modulational instability can be described by analytical formulations such as breather solutions. The Akhmediev breather solution, which is periodic in space and characterized by an amplification factor ranging from 1 to 3, was produced unconsciously in laboratories (e.g., Yuen and Lake 1980). The Kuznetsov-Ma breather solution, which is periodic in time and decreases exponentially in space, was also produced in wave tanks (e.g., Osborne 2010).

Chabchoub et al. (2011) successfully generated a Peregrine breather solution for water waves in a wave tank for the first time, which appears only once in space and time. Shemer and Alperovich (2013) conducted a similar experiment but found a much larger amplification of extreme wave crest with a longer evolutional distance. Besides, Chabchoub also observed a super rogue wave with amplitude up to 5 times the background value in a water wave tank, significantly impacting the study of abnormal waves.

Recalling that the standard modulational instability concerns the instability of a monochromatic wave to long wave perturbations, several questions appear naturally, considering that ocean waves are far from being monochromatic and characterized by a finite spectral width.

Recently, Chabchoub (2016) conducted another extensive experimental study confirming that extreme localizations in an irregular background wave field can be tracked initially back to the exact nonlinear Schrödinger equation (NLS) breather solutions, such as the Peregrine breather. In the more realistic sea state characterized by a unidirectional JONSWAP wave spectrum, Onorato et al. (2005) experimentally confirmed that large amplitude unidirectional waves in random sea states may result from the nonlinear wave focusing. They verified the correspondence of the Benjamin-Feir Index to the extreme wave statistics.

Unidirectional random waves with an initial Gaussian spectrum were produced by Shemer and Sergeeva (2009) in a 300-m-long wave tank. Interestingly, the initial narrow

Gaussian spectrum becomes wider at the early stages of the evolution and then narrower again. Their extensive experiments (Shemer et al. 2010a) demonstrate that the width and shape of the initial spectrum can strongly affect the formation of extreme waves.

Regarding multidirectional waves, Stansberg (1994) reported that nonlinear four-wave interaction might lead to a more significant increase of extreme waves in narrow-banded long-crested waves, while observations in short-crested waves show very little influence of these space-dependent modulational effects. Using the directional JONSWAP wave spectrum, Onorato et al. (2009) and Waseda et al. (2009) revealed that in wave tank experiments, the occurrence of extreme waves is significantly reduced when the directionality broadens.

Toffoli et al. (2011) demonstrated that the number of extreme events depends on the angle between the two interacting systems for the particular case of crossing sea states. Concerning the effect of an opposing current in random wave fields, it has been experimentally investigated by Onorato et al. (2011), revealing the maximum amplitude of rogue waves dependent on the ratio of the current velocity to the wave group velocity.

1.2 Literature Review

1.2.1 Wave Modelling

The mechanisms of generating abnormal or freak waves are various and complicated. Besides the linear superposition of Fourier components with coherent phases and strong wave-current interaction or wave diffraction, it has been established that the modulational instability of waves can explain many features of the dynamics of abnormal waves (Onorato et al. 2001). This instability was first discovered by Lighthill, and the corresponding theory was developed independently by Benjamin and Feir, and by Zakharov.

One simple but useful model, which contains the representation of the modulational instability for the weakly nonlinear evolution of a wave group, is the nonlinear Schrödinger equation (NLS). This equation was first derived by Zakharov with a spectral method and subsequently by Hasimoto and Ono using a multiple-scale technique (see Mei 1989, for a detailed derivation). Even though the NLS equation is just the first step in the hierarchy of complexity of envelope equations, it has considerable merits, such as being integrable and, consequently, a lot of analytical solutions.

One typical example in the explanation of extreme waves with the NLS model is the work of Osborne (2010), where the solution of the one-dimensional nonlinear Schrödinger equation with periodic boundary conditions is written as a "linear" superposition of stable modes, unstable modes, and their mutual nonlinear interactions. The stable modes form a Gaussian background wave field from which the unstable modes occasionally rise and disappear, repeating the process quasi-periodically in time.

Moreover, a quantitative agreement has been observed between the experimental growth rate of the unstable sidebands and the theoretical predictions based on the NLS equation. For the typical ocean sea state described by a JONSWAP spectrum, numerical simulation performed with the NLS equation was first theoretically investigated by Onorato et al. (2001) and then compared with the laboratory experiments by many researchers (e.g., Fedele et al. 2010).

It has been demonstrated that the NLS equation is adequate for a qualitative description of specific global properties of the envelope evolution of unidirectional nonlinear wave groups, such as a local exponential growth in the amplitude of a wave train, but is incapable of capturing more subtle features, for example, the emerging front-tail asymmetry observed in laboratory experiments.

Thus, to enlarge its range of validity, an alternative method is to extend the NLS equation by adding the higher-order nonlinear terms. The first attempt in this direction was made by Roskes, whose numerical simulation displayed two soliton-like envelopes asymmetrically split from an initially symmetric soliton envelope and was in qualitative agreement with the previous observation. However, his envelope equation did not include all the fourth-order terms, and Dysthe (1979) later gave the correct form. Strictly speaking, the Dysthe equation or the modified nonlinear Schrodinger equation (MNLS) is not a fully fourth-order model as its higher-order terms only account for the finite spectral width.

The governing equation can be modified into a spatial form to perform a convenient comparison with experimental results. Such modification of the Dysthe equation was first made by Lo and Mei (1985), who attempted to describe the evolution of the temporally varying wave field downstream along the experimental facility. Further modification of the NLS equation (BMNLS), appropriate for a broader wave spectrum, was made by Trulsen and Dysthe (1996) by adding the higher-order dispersive terms. Moreover, to enhance the MNLS equation with exact linear dispersion (ELMNLS), a pseudo-differential operator can be introduced for the linear part of the BMNLS equation (Trulsen et al. 2000).

As Kit and Shemer (2002) explained, the spatial form of the Dysthe equation can also be easily derived by expanding the dispersion term in the Zakharov equation into a Taylor series. Two versions of the normalized coupled systems of Dysthe equations are derived in their work to describe the spatial evolution of the group envelope. One version was obtained for the amplitude of the velocity potential at the free surface, which is identical to that used by Lo and Mei (1985), while the other one was derived for the amplitude of the surface elevation. As Trulsen and Dysthe (1997) stated, these two versions will become consistent if the first harmonic of the surface displacement is of primary interest for comparison with experimental observations.

It has been demonstrated that for the simple unidirectional wave packet with an initial Gaussian-shaped envelope, numerical computations based on the Dysthe model indeed provide a good agreement with experiments and exhibit the front-tail asymmetry (Lo and Mei 1985; Shemer et al. 2002, 2010a). Comparison with more complicated sea states generated in the wave basin (Onorato et al. 2005) has illustrated that the spatial variation of the coefficient of kurtosis, a potential indication for the generation of extreme events, can be reproduced well by the numerical simulation of Dysthe model. Nevertheless, the validity of the Dysthe model is still limited to a specific spatial domain in the wave tank, and a noticeable discrepancy can be observed in the case of numerical simulation of shorter-length waves.

To include the influence of directional spreading, Toffoli et al. (2010a) detailedly investigated the discrepancy of three-dimensional waves with various degrees of directionality and the difference of simulations of spatial and temporal versions of the 3D Dysthe model. For short-crested waves, the deviation from Gaussian statistics is only due to bound wave contributions, while free waves preserve normal statistics despite the third-order nonlinear evolution. On the foundation of a large set of numerical simulations, Gramstad and Trulsen (2007) performed a more systematic analysis of the effect of directionality, revealing how freak waves in deep water depend on the group and crest lengths for a fixed steepness. The corresponding experimental investigation was first made by Stansberg (1994), who casted some light on the effects of directionality on the fourth-order moment of the surface elevation. More recent experimental results were reported by Waseda (2005) and Onorato et al. (2009), indicating that the occurrence of extreme waves can be significantly reduced when the directionality broadens.

Another very effective method to investigate abnormal waves is the quasi-determinism (QD) theory of Boccotti (2014), as demonstrated in comparison with the mechanically generated New Year wave (Petrova et al. 2011). Meanwhile, somewhat more general alternatives to NLS-type models are the so-called Zakharov equations and the closely related mode-coupling approaches. To include the higher-order (quintet) nonlinear interaction, the fourth-order Zakharov equation was also derived by Stiassnie and Shemer. However, as Shemer et al. (2002) pointed out, for wave groups with an initial narrow spectrum, the numerical simulation of the Zakharov equation makes no significant difference from that of Dysthe equation, which requires substantially less computational resources.

By extending the mode-coupling idea, a more direct numerical approach, known as the high-order spectral method (HOS) for gravity waves, was proposed respectively by West et al. (1987) and Dommermuth and Yue (1987). This method accounts for nonlinear interactions up to a specified order in wave steepness, provided the Taylor series expansion of the velocity potential on the free surface around the mean waterline remains valid. In this respect, the HOS method's numerical simulation is not fully nonlinear or exact. However, due to its fast convergence and high computational efficiency, the original and later enhanced versions of the HOS method have been widely applied in the simulation of 2D and 3D ocean waves, particularly in the research of extreme wave generation (e.g., Ducrozet et al. 2007; Toffoli et al. 2008a, 2010a; Slunyaev et al. 2014).

Recently, it has been found that a more straightforward and more precise, fully nonlinear approach can be constructed on the basis of nonstationary conformal mapping, which has been formulated long before it was used for numerical integration. Chalikov and Sheinin (1998, also referred to as CS) described the detailed numerical scheme for arbitrary depth of water wave and its validation. For a nonstationary problem, the CS numerical approach allows rewriting the principal equations of potential flow with a free surface in a surface-following coordinate system. The Laplace equation retains its form while the boundary conditions are coordinate surfaces in the new coordinate system. The velocity potential in the entire domain can be represented by its Fourier expansion coefficients on the free surface, and the hydrodynamic system, without any simplification, is expressed by two relatively simple evolutionary equations that can be solved with a straightforward numerical method.

Many highly or fully nonlinear models, which can not be listed exhaustively herein, have been developed to study various complex wave phenomena in the past decades. As reviewed by Dias and Kharif (1999), the common drawback of the existing fully nonlinear methods is the slow computational speed. Motivated by seeking a fast Laplace equation solver, Clamond and Grue (2001) developed a novel rapid method employing integral equations while its computational efficiency was only validated by relatively simple wave fields.

It has to be admitted that although the Navier–Stokes equation (or large eddy simulation equation) requires much larger computational resources, some issues, such as two-phase flow and wave breaking, beyond the scope of potential theory, have to be researched with such kinds of models. For example, Iafrati et al. (2013) investigated the dynamics of the modulational instability of free surface waves and its contribution to the interaction between ocean and atmosphere, and Perić et al. (2015) studied the Peregrine breather dynamics up to the initial stages of wave breaking.

1.2.2 Wave Statistics

In a linear random sea state, the relatively large waves are regular in the sense that they do not display any secondary maxima or minima in a wave period. Thus, the expected shape of large surface displacements and the distributions of wave amplitudes can be described reasonably well by Gaussian statistics and the Rayleigh model, respectively (Boccotti 2014). Furthermore, the additional assumption that the wave height equals twice the wave amplitude makes the Rayleigh distribution easily applicable to wave height.

However, the ocean waves are nonlinear, represented by higher, more peaked crests and shallower, more rounded troughs. Consequently, the distribution of surface displacements tends to deviate from the Gaussian form with positive skewness, and the linear Rayleigh models are usually unsuitable for this case. Hence, proposing a more accurate model is theoretically important and practical in ocean engineering, as it can predict wave forces and structural responses.

Concerning the surface elevation, the solution was first given by Longuet-Higgins, who approximated its distribution in a Gram–Charlier type series, type A, in the form of Edgeworth. The Gram–Charlier representation is valid in the deepwater wave. It can be extended further to the shallow water case, even though sometimes it will become negative over the range of large negative amplitudes. Later, this drawback is overcome by some other researchers, who derive a new nonlinear distribution from a Pearson system of distributions.

As for the wave amplitude, Tayfun explored a Stokes-like model and proposed a simple theoretical expression to describe the distribution of nonlinear wave amplitudes in narrowband waves. After that, numerous nonlinear wave amplitude models began to appear, derived with similar methods (Forristall 2000; Arena and Fedele 2002).

To sum up, some models use nonlinear transforms of Gaussian variables analogous to Gram–Charlier type expansions or fit observational data to a theoretical distribution defined in terms of parameters which include second-order nonlinear effect; some represent extensions of the narrowband approximations to shallow-water waves or make some heuristic modifications of the Tayfun model; and others require intricate analysis and numerical computations (Fedele and Tayfun 2009). Most of them can provide valuable insight and work well with various degrees of success in describing the distribution of crests and troughs of giant waves.

Wave heights are independent of second-order bound wave effects because, in the second order, the crest and trough are shifted equally upwards, as in the deterministic Stokes waves and nearly so in the narrowband waves. Hence, it is expected that they can be fitted very well by the Rayleigh law even though the higher-order bound wave effects, which are pretty small, are not included.

However, the frequency of occurrence of relatively rare and unusually large oceanic waves commonly referred to as abnormal, rogue or freak waves significantly exceeds the theoretical predictions based on all second-order models, especially in the case of long-crested, steep and narrowband waves (Tayfun and Fedele 2007; Cherneva et al. 2009). In fact, to the third order in wave steepness ε, there is a substantial change in the description of water waves. While the bound wave modes are still present, resonances (or quasi-resonances) are also possible among free waves that consequently can cause a local exponential growth in wave amplitudes within a time scale of $1/\varepsilon^2$ wave periods (Janssen 2003). The mechanism involved is a generalization of the Benjamin–Feir instability or modulational instability, which formally applies only to a Stokes wave and a small perturbation.

In a recent study, the third-order nonlinearity, represented by the fourth-order cumulants, has been successfully applied in the statistics of abnormal waves, of which the probability is described reasonably well by the theoretical approximation based on Gram–Charlier (GC) expansions (Mori and Janssen 2006; Tayfun and Fedele 2007). The deviation from the Gaussian structure is mainly attributed to the modulational instability in the wave train, which can be described quantitatively through some critical parameters.

Take the third-order NB-GC model as an example (Tayfun and Fedele 2007); the fourth-order normalized joint cumulants are responsible for the large amplitude events and the increased probability of occurrence of abnormal waves as confirmed in a series of studies (e.g., Onorato et al. 2006; Mori et al. 2007; Petrova et al. 2007; Shemer et al. 2010a). Besides, to consider the influence of the bound wave effect on the wave amplitude, a second-order distribution has replaced the leading Rayleigh term.

Finally, it has to be pointed out that the directional spreading can significantly decrease the effect of modulational instability and, thus, the formation of abnormal waves. Therefore, the second-order models typically can work very well in the case of short-crested waves (Onorato et al. 2009).

1.3 Outline of Chapters

Except for the introduction chapter, the other chapters can be simply categorized into two parts according to the issues addressed in this book.

The first part is Chap. 2, where the laboratory experiments carried out in Marintek, Norway, and in CEHIPAR, Spain, are comparatively studied together with both spectral and statistical methods to verify and validate the data. Due to the different model scales used in the experiments, only the time series of three cases are compared directly on some specified locations that, on a full scale, have the same distance from the wave-maker. A comparison has been made between the main sea state parameters, the power spectral density, joint distribution of wave heights and periods, and marginal distributions of wave heights and wave periods, respectively. Meanwhile, some observations are also compared with the corresponding theoretical model predications. Furthermore, the directional spreading effect on the formation of extreme waves is also discussed.

The laboratory experiments performed in the Shandong Provincial Key Laboratory of Ocean Engineering are mainly used to analyze the dynamic pressure variation under the crest of nonlinear waves. The vertical pressure profile of individual waves with the same crest height is investigated in regular and irregular sea states, and the influence of different types of nonlinearity on dynamic pressure is discussed in both linear and nonlinear focused waves.

The second part, from Chaps. 3–8, is mainly related to the numerical simulations of extreme waves with different numerical models, starting from the most straightforward cubic nonlinear Schrödinger equation to the fully nonlinear CS model. These governing equations also theoretically study the spatial and temporal evolutions of typical nonlinear wave groups with the same initial wave condition. Moreover, laboratory experiments run in the offshore wave basins validated in the first part are systematically compared with a large set of numerical simulations performed by these numerical models. Apart from the high-order statistical moments such as the coefficients of skewness and kurtosis, the distributions of surface elevation, wave crest, wave trough, wave height and wave period are also studied in sequence.

Moreover, the influence of irregular background waves on triggering abnormal waves is investigated from a statistical point of view. Meanwhile, for a better understanding of the higher-order nonlinearity of Peregrine breather dynamics, the corresponding evolutionary tendency in regular background waves is also given under the control of a fully nonlinear HOS model. The effect of initial condition uncertainty on the profile of maximum wave is considered in two typical short-term sea states. The associated uncertainty is investigated using the Monte Carlo Simulation method, and the maximum wave is extracted from the numerical simulation performed with the fully nonlinear CS model that was validated in the previous chapter. Six parameters are defined to describe the property of the profile of the maximum wave and analysed with statistical methods.

2 Laboratory Experiments Related to Extreme Waves

2.1 Basic Theory

2.1.1 Carrier Wave Envelope Representation

Consider the deepwater wave with the narrowband condition and let $\eta(t)$ represent the surface elevation at a fixed point in time t. If $S(\omega)$ is a frequency spectrum of $\eta(t)$ and $m_i = \int_0^\infty \omega^i S(\omega)d\omega$ are the spectral moments, then a frequency $\omega_{01} = m_1 / m_0$ can be defined. The surface elevation $\eta(t)$ and its Hilbert transform $\hat{\eta}(t)$ define a wave envelope with the amplitude $\rho = \left[\eta^2(t) + \hat{\eta}^2(t)\right]^{1/2}$ and the total phase $\varphi(t) = \arctan\left[\hat{\eta}(t) / \eta(t)\right] = \omega_{01}t + \phi(t)$ that is increasing in time. Now the surface elevation can be represented as a process with a fixed carrier frequency ω_{01}, modulated by a complex wave envelope with amplitude ρ and phase φ where ρ and φ are real and time-varying random variables.

$$\eta(t) = \mathrm{Re}\left\{\rho e^{i\phi} \exp(i\omega_{01}t)\right\}. \tag{2.1}$$

Additionally, define the mean wave period corresponding to the carrier frequency as $\overline{\tau} = T_{01} = 2\pi / \omega_{01} = 2\pi m_0 / m_1$, the spectrum width $\nu = \left[\left(m_0 m_2 / m_1^2\right) - 1\right]^{1/2}$ and the wave steepness $\varepsilon = H_s \omega_p^2 / 2g$, where H_s is the significant wave height, ω_p the spectral peak frequency and g is the acceleration of gravity.

2.1.2 Longuet-Higgins Joint Distribution

If the amplitude ρ and the phase φ in Eq. (2.1) are slowly varying functions of time, then the maxima and minima of $\eta(t)$ lie near to the envelope function, and the energy spectrum is concentrated in a small frequency interval around ω_{01}. Under this assumption,

H. Zhang and C. Guedes Soares, *Numerical Modelling of Extreme Waves*, Synthesis Lectures on Ocean Systems Engineering, https://doi.org/10.1007/978-3-031-77084-5_2

the wave height H could be treated as $H = 2a = 2\rho$, i.e., twice the wave amplitude a, which is equal to the magnitude of the envelope function ρ. The wave period can also be expressed by the rate of change of the total phase as $\tau = 2\pi / \dot{\phi}$ with $\dot{\phi} = \omega_{01} + \dot{\varphi}$. As a result, a joint probability density $p(\rho, \dot{\varphi})$ was deduced by Longuet-Higgins. Later, a further procedure is taken as below to correct the apparent deviation in the wave period:

$$R = \rho \Big/ (2m_0)^{1/2} = H \Big/ (8m_0)^{1/2}, \tag{2.2}$$

$$T = \tau / \overline{\tau}. \tag{2.3}$$

Thus, a new, improved joint distribution of wave heights and periods is derived (Longuet-Higgins, 1983).

$$p(R, T) = \frac{2L(\nu)}{\nu\sqrt{\pi}} \frac{R^2}{T^2} e^{-R^2\left[1+(1-1/T)^2/\nu^2\right]}. \tag{2.4}$$

Here $L(\nu)$ is a factor introduced to normalize the total probability to unity. After integration of Eq. (2.4) over positive values of R and T the normalizing factor is given by

$$L(\nu) = \frac{2\sqrt{1+\nu^2}}{1+\sqrt{1+\nu^2}}. \tag{2.5}$$

For a narrow spectrum that is defined as $\nu^2 \le 0.36$ or $\nu \le 0.6$, $L(\nu) \approx 1 + \nu^2 / 4$, almost equal to unity. Moreover, the maximum value of $p(R, T)$ on the distribution is

$$p(R, T)_{max} = (2L/\pi^{1/2} e)(1+\nu^2)/\nu. \tag{2.6}$$

The distribution Eq. (2.4) implies that the location of the mode of the joint distribution always satisfies the condition $T < 1$, and the new theoretical model is asymmetric with respect to the period T. Only for an unreal narrow spectrum of width $\nu \le 0.1$ the distribution approaches to being symmetrical about the mean wave period.

Integration of Eq. (2.4) over $T \in (0, \infty)$ gives the marginal density of wave heights (Longuet-Higgins, 1983)

$$p(R) = \int_0^\infty p(R, T)\, dT = 2R \exp(-R^2) \left[L(\nu) \frac{1}{\sqrt{\pi}} \int_{-\infty}^{R/\nu} e^{-\beta^2} d\beta \right]. \tag{2.7}$$

where $\beta = R(1 - 1/T)/\nu$. The marginal distribution of heights comprises a Rayleigh part and a correction that only depends on the spectrum width ν.

Similarly, integrating Eq. (2.4) over $R \in (0, \infty)$ gives the marginal distribution of wave periods.

$$p(T) = \int_0^\infty p(R,T)\,dR = \frac{L(\nu)}{2\nu T^2}\left[1 + \left(1 - 1/T\right)^2/\nu^2\right]^{-3/2}. \tag{2.8}$$

As expected, the distribution $p(T)$ is not symmetrical, and its mode always has a value of T that is less than unity. As the spectral width ν increases, the mode shifts to a relatively small period.

2.1.3 Modified Longuet-Higgins Joint Distribution I

It is widely accepted that the JONSWAP spectrum characterizes the ocean waves and that the narrowband condition can be strictly satisfied for the unidirectional waves mechanically generated in the laboratory. Consequently, the peak frequency ω_p is very close to the mean frequency ω_{01}, and most of the energy is concentrated around the peak frequency in the spectrum. In this sense, it is reasonable to use ω_p to work as the carrier frequency as what is widely done in wave modelling. Moreover, the mean frequency ω_{01} calculated directly from a bimodal wave spectrum is no longer the carrier frequency in mixed sea states. In contrast, each peak frequency is more suitable for working as the carrier frequency of wind waves and swell separately in this case. To derive the alternative Longuet-Higgins joint distribution, the surface elevation is now represented in a new form:

$$\eta(t) = \mathrm{Re}\left\{\rho_1 e^{i\phi_1} \exp\left(i\omega_p t\right)\right\}. \tag{2.9}$$

Considering that the major derivation procedures are almost identical to those for the Longuet-Higgins joint distribution, only the critical parts are presented below. Since the carrier frequency has been changed, it is necessary to define a new kind of spectral moment to replace the original central moment.

$$\tilde{\mu}_n = \int_0^\infty \left(\omega - \omega_p\right)^n S(\omega)\,d\omega. \tag{2.10}$$

Thus, the new first-order spectral moment is given by $\tilde{\mu}_1 = m_1 - \omega_p m_0$. For a narrowband spectrum, the mean frequency will approach the peak frequency with increasing steepness and thus $\tilde{\mu}_1 \to 0$. Consequently, the four normally distributed random variables utilized in the derivation process can still be treated as independent variables because the correlation matrix is approximately diagonal. Furthermore, it has been empirically found that those representative wave periods are interrelated. Thus, it is feasible to set up a relationship between the mean and the peak period to retain the wave spectral width parameter ν in the newly modified joint distribution.

$$T_{01} = \beta T_p. \tag{2.11}$$

The wave periods will be normalized by the peak period rather than the mean period, while the normalized parameter of wave heights will still be the same as before. After some algebraic operations, the new joint distribution, also named the modified Longuet-Higgins joint distribution I, is:

$$p(R, T_1) = \frac{2\alpha C(\alpha) R^2}{\sqrt{\pi} T_1^2} \exp\left\{ -R^2 \left[1 + \alpha^2 \left(1 - \frac{1}{T_1} \right)^2 \right] \right\}, \tag{2.12}$$

where

$$C(\alpha) = 2\sqrt{1+\alpha^2} \Big/ \left(\sqrt{1+\alpha^2} + \alpha \right), \tag{2.13}$$

$$\alpha = \beta \Big/ \sqrt{(\beta - 1)^2 + \nu^2}. \tag{2.14}$$

This modified joint distribution depends on one parameter α, determined by the wave spectral width parameter ν and the ratio of mean period to peak period β. The new position of the mode can be found in the same way as before:

$$\left. \begin{aligned} R &= \alpha \Big/ \sqrt{1+\alpha^2} \\ T_1 &= \alpha^2 \big/ \left(1+\alpha^2\right) \end{aligned} \right\}. \tag{2.15}$$

The probability of density mode is

$$p_{max} = \left(2C/e\sqrt{\pi} \right) \left(1 + \alpha^2 \right) \big/ \alpha. \tag{2.16}$$

Furthermore, the modified marginal distributions of wave heights and wave periods are given by

$$p(R) = R \exp\left(-R^2\right) C(\alpha) \left[1 + erf(\alpha R) \right], \tag{2.17}$$

$$p(T_1) = \frac{\alpha C(\alpha)}{2T_1^2} \left[1 + \alpha^2 \left(1 - \frac{1}{T_1} \right)^2 \right]^{-3/2}, \tag{2.18}$$

where the error function is

$$\mathrm{erf}(x) = \frac{2}{\sqrt{\pi}} \int_0^x \exp\left(-t^2\right) dt, \tag{2.19}$$

2.1.4 Modified Longuet-Higgins Joint Distribution II

Some terms have been omitted in the derivation of Longuet-Higgins joint distribution I due to the assumption that $\tilde{\mu}_1 \to 0$. The merits of taking this limit will be seen in the later comparison with laboratory observations. Now, proposing a more exact joint distribution without any hypothesis is possible. After much more tedious algebraic operations, the modified Longuet-Higgins joint distribution II is obtained.

$$p(R, T_1) = \frac{2\beta L(\nu)R^2}{\nu\sqrt{\pi}T_1^2} \exp\left\{-R^2\left[1 + \frac{1}{\nu^2}\left(1 - \frac{\beta}{T_1}\right)^2\right]\right\}. \tag{2.20}$$

The position of the mode is also found in the following expressions

$$\left.\begin{array}{l} R = 1\Big/\sqrt{1+\nu^2} \\ T_1 = \beta\big/1+\nu^2 \end{array}\right\}. \tag{2.21}$$

and the value of $p(R, T_1)$ at this point is

$$p_{max} = \left(2L/\beta e\sqrt{\pi}\right)\left(1+\nu^2\right)\big/\nu. \tag{2.22}$$

As anticipated, the marginal distribution of wave heights is still the same as that expressed in Eq. (2.7) and will not be repeated here. The modified marginal function of wave periods is derived in the same manner as before, leading to:

$$p(T_1) = \frac{\beta L(\nu)}{2\nu T_1^2}\left[1 + \frac{1}{\nu^2}\left(1 - \frac{\beta}{T_1}\right)^2\right]^{-3/2}. \tag{2.23}$$

It is noted that the original and modified marginal distributions of wave heights expressed in Eqs. (2.7) and (2.17) are both nearly Rayleigh distributed. However, they have a correction factor that reduces the probability of small amplitude waves and increases the probability of waves near the mode, shifting the mode slightly to higher values. The correction has an exponentially small effect on the tail of the Rayleigh distribution, but that part is of great interest in ocean engineering.

2.1.5 Nonlinear Models of Wave Height Distribution

If the wave height is expressed as the double local amplitude of the envelope in a narrowband Gaussian sea state, the exceedance distribution of wave height can be described in the scaled Rayleigh form as

$$E_{Ray}(R) = \exp(-R^2). \tag{2.24}$$

As discussed by Tayfun and Fedele (2007), the correction due to third-order nonlinear interaction can be described by the Gram–Charlier series expansions. Besides, Mori and Janssen (2006) also proposed a modified Edgeworth–Rayleigh (MER) distribution, where the coefficient of kurtosis λ_{40} indicates the influence of third-order nonlinearity. To keep consistent with the former wave height distribution, the MER distribution is scaled by the root-mean-square wave height and thus, the original formula of this model is changed into a new one:

$$E_{MER}(R) = E_{Ray}(R)\left[1 + \lambda_{40}R^2\left(R^2 - 2\right)/6\right]. \tag{2.25}$$

2.2 Unidirectional Wave Experiments in Two Wave Basins

2.2.1 Marintek Wave Basin

As shown in Fig. 2.1, data in Marintek, the dimensions of which are 80 m long and 50 m wide, were obtained during a series of laboratory experiments run in Trondheim in 1999 (Cherneva et al. 2009). The wave surface elevations are measured by ten capacitance wave gauges deployed along the centreline of the basin with a water depth of 2 m. The length scale of the experiments is 1:50, and the gauge locations are listed in the model scale in Table 2.1.

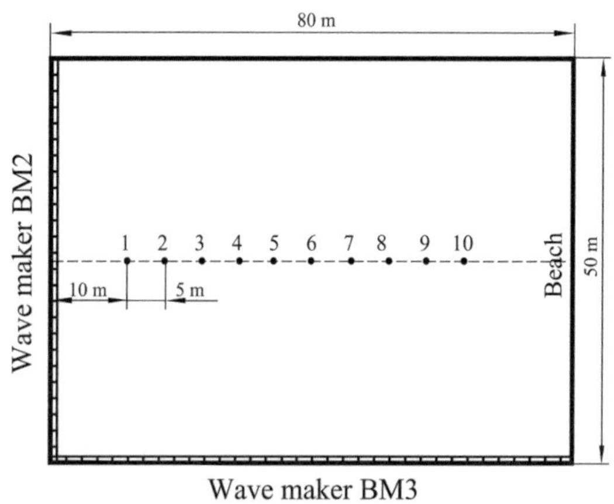

Fig. 2.1 Layout of Marintek wave basin and gauge locations

Table 2.1 Gauge locations in Marintek wave basin in model scale

Gauge	1	2	3	4	5	6	7	8	9	10
Dist. (m)	10	15	20	25	30	35	40	45	50	55

Table 2.2 Initial parameters of experiments in Marintek

Sea State	No.	H_s(m)	T_p (s)	ε	γ	*BFI*	Duration (h)
Smooth	8201	3.5	10	0.070	3.0	0.50	3
Moderate	8202	7.0	10	0.141	3.0	0.99	3
	8241–45	3.5	7.0	0.144	3.0	1.02	3
Severe	8219	9.0	10	0.181	3.0	1.28	3

The waves generated at the wavemaker are unidirectional. Each experiment uses different sets of random phases, and the variance of amplitudes is such that the spectrum of the waves generated at the wavemaker represents a JONSWAP spectrum characterized by the peak enhancement factor $\gamma = 3$ and the Philips parameter α_1 depending on the desired significant wave height. The laboratory facility, the setup of experiments and some characteristics of these produced wave series have been given in detail by Cherneva et al. (2009, 2013) and Cherneva and Guedes Soares (2011).

Note that *BFI* represents the Benjamin–Feir Index, of which the definition and meaning will be given in the following chapter. Each time series lasts approximately 3 h on a full scale. All initial parameters of the four different experiments are listed in Table 2.2, and 8241–45 is a set of records from five independent tests run under the same sea state.

2.2.2 CEHIPAR Wave Basin

The wave basin in CEHIPAR, Spain, is 152 m long, 30 m wide and 5 m deep, as sketched in Fig. 2.2. The wavemaker is located at one of the 30 m sides. The waves are produced by 60 flaps with independent motion. On the opposite side of the wavemaker, there is a wave beach that absorbs the incident wave energy. The wavemaker can produce long and short-crested sea states with up to 0.4 m significant wave heights and standard or arbitrary shape spectra.

The length scale of the laboratory experiments examined here is 1:40. The waves generated for this study are long-crested. They are registered by six capacitance wave gauges situated in the midline of the basin. Each experiment with the same initial condition is carried out two times. To increase the number of measurement locations, the six gauges are moved downstream by 10 m, and the same realizations are repeated. The detailed gauge locations are listed in Table 2.3 and plotted in Fig. 2.3.

There are 23 cases in the laboratory experiments with different initial wave parameters listed in full scale in Table 2.4. The duration of each recorded time series is nearly half an hour on a model scale with a sampling frequency of 100 Hz.

According to the initial wave steepness ε, the cases in Tables 2.4 and 2.2 can be roughly categorized into three groups: the smooth, moderate and severe sea states, and will be represented by different symbols in the later analysis.

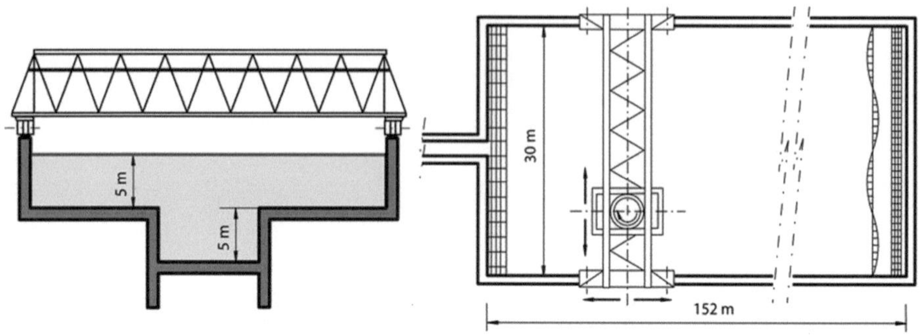

Fig. 2.2 Layout of CEHIPAR wave basin

Table 2.3 Gauge locations in CEHIPAR wave basin in model scale

Gauge	1	2	3	4	5	6
Part A (m)	20	40	60	80	100	120
Part B (m)	30	50	70	90	110	130

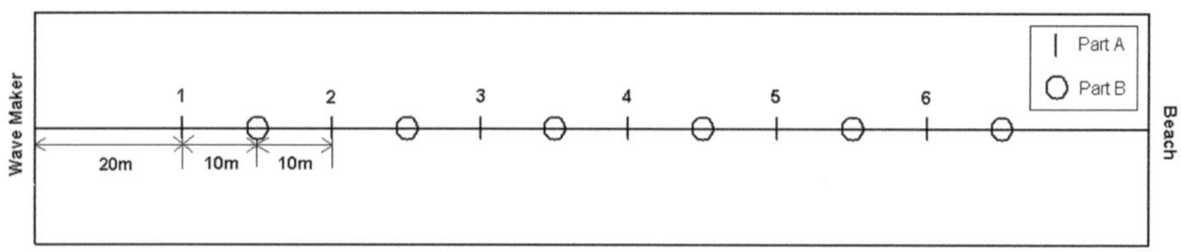

Fig. 2.3 Location of wave gauges in CEHIPAR

Table 2.4 Initial parameters of experiments in CEHIPAR

Case	H_s (m)	T_p (s)	ε	Case	H_s (m)	T_p (s)	ε	Case	H_s (m)	T_p (s)	ε
	Smooth				Moderate				Severe		
1	5	14	0.051	8	3	7	0.123	16	4	7	0.164
2	6	14	0.062	9	9	12	0.126	17	12	12	0.168
3	6	13	0.071	10	8	11	0.133	18	7	9	0.174
4	8	14	0.082	11	7	10	0.141	19	9	10	0.181
5	12	16	0.094	12	3.5	7	0.144	20	11	11	0.183
6	5	10	0.101	13	6	9	0.149	21	6	8	0.188
7	12	15	0.107	14	11	12	0.154	22	12	11	0.199
				15	8	10	0.161	23	5	7	0.205

Table 2.5 Initial full-scale sea state parameters of experiments in comparison

H_s(m)	T_p(s)	Steepness	Marintek (Exp. No.)	CEHIPAR (Exp. No.)
7.0	10	0.141	8202	11
3.5	7	0.143	8241–8245	12
9.0	10	0.181	8219	19

Table 2.6 Gauge locations that have approximately equal distance to the wavemaker

Marintek Gauge	Dist. (m)	CEHIPAR Gauge	Dist. (m)
4	1250	1	1210.16
7	2000	2	2000.12
10	2750	3	2798.44

2.2.3 Experimental Data for Comparative Analysis

Duc to the different model scales in the laboratory experiments made in Marintek and CEHIPAR basins, only 3 cases, listed in Table 2.5, have the same initial sea state conditions. Three records in each case can be directly compared and are measured approximately at the same distance from the wavemaker (sec Tablc 2.6 in full scale).

There is little difference between the distances, as presented in the first and third rows of Table 2.6. However, the discrepancy can be ignored compared with the dominant wavelength, which is 156.13 m, corresponding to the peak period of 10 s.

Since the initial wave steepness is largcr than 0.14, these wave series are qualified enough to check the measurement accuracy of nonlinear wave series. Not only the spectral but also the statistical methods will be used to analyze these series. For clarity, some theoretical prediction models will be compared as well.

2.3 Verification and Validation of Extreme Wave Experiments

2.3.1 Sea State Parameters

The spatial variations of zero spectral moments and mean wave periods corresponding to the three sea states in Table 2.5 are plotted in Fig. 2.4, where the Marintek data are marked by circles and the CEHIPAR data are represented by triangles. All figures in this section will adopt the same symbols to identify the results of the two wave basins.

Although in the moderate sea states, the zero spectral moments shown in Fig. 2.4a and b give a tiny discrepancy in the estimated values, in the severe sea state, they present the same spatial variation along the wave basins, i.e., Fig. 2.4c. Therefore, it can be concluded that the change of wave energy in the two wave basins is almost the same before 3000 m.

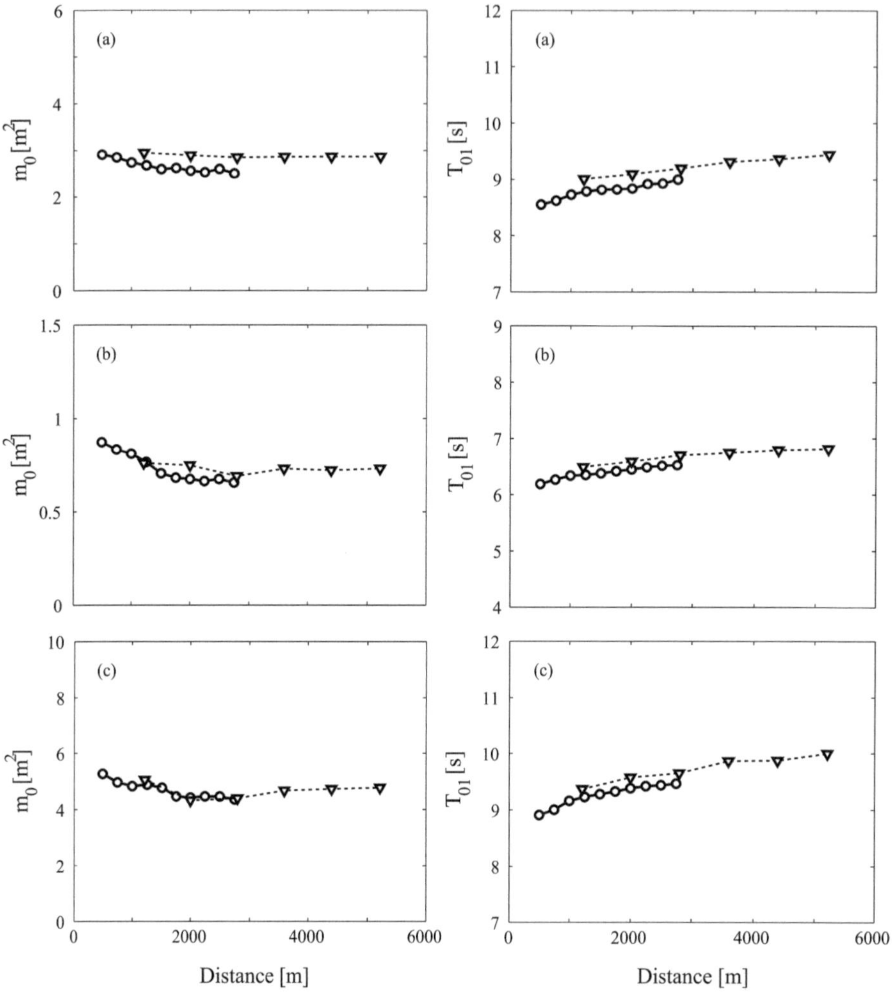

Fig. 2.4 Spatial variations of zero spectral moments m_0 and mean wave periods T_{01} in Marintek (circles connected by a solid line) and in CEHIPAR (triangles connected by a dashed line). **a** $H_s = 7$ m, $T_p = 10$ s; **b** $H_s = 3.5$ m, $T_p = 7$ s; **c** $H_s = 9$ m, $T_p = 10$ s

In all cases, the mean wave period increases downstream in the two wave basins and displays a linear relationship with the distance to the wavemaker. The mean wave period in CEHIPAR is always larger than that in Marintek. This discrepancy can be ignored because the largest difference is not more than 6%.

The other important sea state parameters are compared in Table 2.7. Both wave spectral width parameter ν and wave steepness ε present an overall decaying tendency along the wave basins before 3000 m for all cases. It is very interesting to note that the spectral

Table 2.7 Comparisons of some dimensionless sea state parameters

Dist. (m)	Case	ν	ε	λ_{30}	λ_{40}
1250	Marintek 8202	0.317	0.134	0.188	0.272
	CEHIPAR 11	0.287	0.134	0.123	0.342
	Marintek 8241–45	0.267	0.136	0.216	0.525
	CEHIPAR 12	0.266	0.132	0.193	0.233
	Marintek 8219	0.289	0.181	0.221	0.573
	CEHIPAR 19	0.275	0.175	0.172	0.063
2000	Marintek 8202	0.304	0.117	0.211	0.394
	CEHIPAR 11	0.276	0.128	0.212	0.755
	Marintek 8241–45	0.261	0.133	0.225	0.680
	CEHIPAR 12	0.249	0.123	0.205	0.369
	Marintek 8219	0.282	0.149	0.220	0.584
	CEHIPAR 19	0.265	0.156	0.216	0.259
2750	Marintek 8202	0.285	0.116	0.132	0.553
	CEHIPAR 11	0.275	0.122	0.235	0.957
	Marintek 8241–45	0.254	0.126	0.214	0.663
	CEHIPAR 12	0.251	0.119	0.203	0.695
	Marintek 8219	0.281	0.152	0.223	0.629
	CEHIPAR 19	0.272	0.157	0.268	0.759

width in Marintek is always a little larger than that in CEHIPAR, which may be attributed to the different aspect ratios of wave basins.

There is no distinct difference between the vertical asymmetry of the wave profile provoked by the second-order bound wave effects in the two basins, as the discrepancy between the coefficient of skewness λ_{30} estimated along the basins is in a reasonable range in most cases.

However, the kurtosis coefficients vary largely along the wave basins and sometimes show a pronounced difference between Marintek and CEHIPAR. This means that the quasi-resonant third-order nonlinear interaction due to nonlinear instability is complex and unstable, and it is strongly influenced by the initial condition synthesised in the wavemaker. Moreover, the different lengths and number of time series between the two wave basins can also lead to a diverse evaluation of the coefficient of kurtosis and an influence on the tail of wave height distributions.

Although they are not presented here, it has to be mentioned that there are some relationships among these four sea state parameters, particularly between the wave steepness and the coefficients of skewness and kurtosis. For a unidirectional weakly nonlinear narrowband wave train, up to second order, the analytical formulae are given by Mori and Janssen (2006). There is no doubt about the linear relationship between the coefficient of skewness and the wave steepness. Nevertheless, the theoretical quadratic function largely underestimates the coefficient of kurtosis for a given steepness in both wave basins. The conclusions are also confirmed by comparing a towing tank and an offshore basin (Cherneva and Guedes Soares, 2011).

2.3.2 Wave Spectrum

Since the significant sections of wave spectra derived in the two wave basins are almost in agreement with each other in the same sea state, as an illustration, only the spatial variation of wave spectral density in the case with $H_s = 7$ m and $T_p = 10$ s is depicted in full scale in Fig. 2.5.

The spatial evolution of wave spectra is not significant even in the moderate sea state (Cherneva et al. 2013). However, in the very severe sea state, the high-frequency range of the wave spectrum, i.e., $\omega > \omega_p$ tends to behave more clearly from ω^{-5} to ω^{-6}. In Fig. 2.5, as indicated by the slightly decreased slope of the straight line ($\omega > \omega_p$), the wave spectrum becomes a little narrower downstream in the wave basin, which is also revealed by the variation of wave spectral width parameter ν in Table 2.7. It can be argued

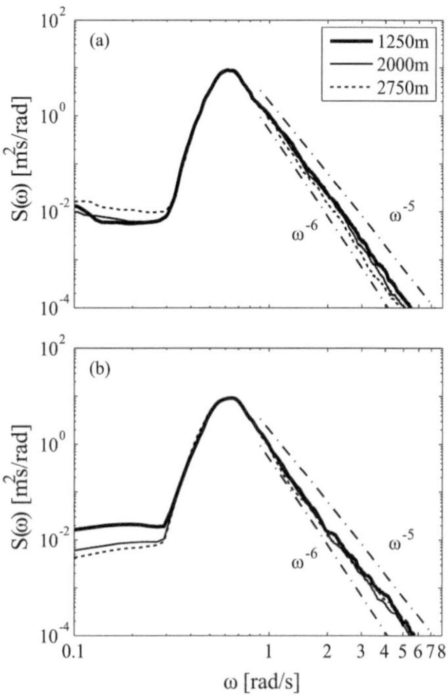

Fig. 2.5 Spatial variation of wave spectrum in moderate sea state. **a** Marintek; **b** CEHIPAR

that this is mainly caused by wave breaking because the viscous dissipation is insignificant in the wave basin (Fedele et al. 2010).

Concerning the high-frequency range of the wave spectrum, analysis based on data sets from different wave fields reveals that the exponents for the tails show less variation in wavenumber spectra than in frequency spectra. A simple model spectrum with a tail $k^{-3.5}$ may be applicable both in deep and shallow water. Nevertheless, the transition between power law models in the high-frequency range exists. The uncertainty of the high-frequency tail slope results from the intrinsically random variability of wind-generated waves and the estimation procedures of the wave spectra.

2.3.3 Joint Distribution of Wave Heights and Wave Periods

For the economy of space and according to the above-presented features of the investigated time series, in the following part of this section, it is feasible and necessary to demonstrate the comparisons of distributions in detail at a specified location, such as 2000 m away from the wavemaker, for all cases.

As seen in Fig. 2.6, the joint distributions of measured wave heights and periods are compared between the two wave basins for the three cases. Moreover, the observed statistics are also compared with the theoretical model predictions of Longuet-Higgins (1983) represented in Eq. (2.4).

The contours take the value $(0.99, 0.9, 0.7, 0.5, 0.3, 0.1) \times p_t$ (or p_o) from the centre contour outwards for both theoretical distribution (solid line) and observed result (coloured area) where p_t (or p_o) means the theoretical (or observed) value of density mode. Firstly, it is possible to conclude that the joint distributions of the scaled wave heights and periods checked in the two wave basins are compatible. Secondly, the theoretical joint distribution of Longuet-Higgins (1983) is wider in comparison with the observed result. In addition, the position of the mode is always shifted to the right side of the theoretical one, and the deviation is enlarged as the wave steepness increases in both wave basins.

Comparing the six wave spectral width parameters ν in 2000 m presented in Table 2.7, it can be inferred that the relative variations of the six theoretical joint distributions are not significant because the theoretical model of Longuet-Higgins (1983) is only dependent on the parameter of wave spectral width.

However, the discrepancies among experiments in the three cases are noticeable, meaning that the joint distribution depends on the wave spectral width and may be related to other sea state parameters, such as wave steepness. In a word, the Longuet-Higgins (1983) joint distribution should be further improved, for instance, to include the higher-order nonlinear effect, especially in the severe sea state. Such an issue will be addressed in Sect. 2.4.

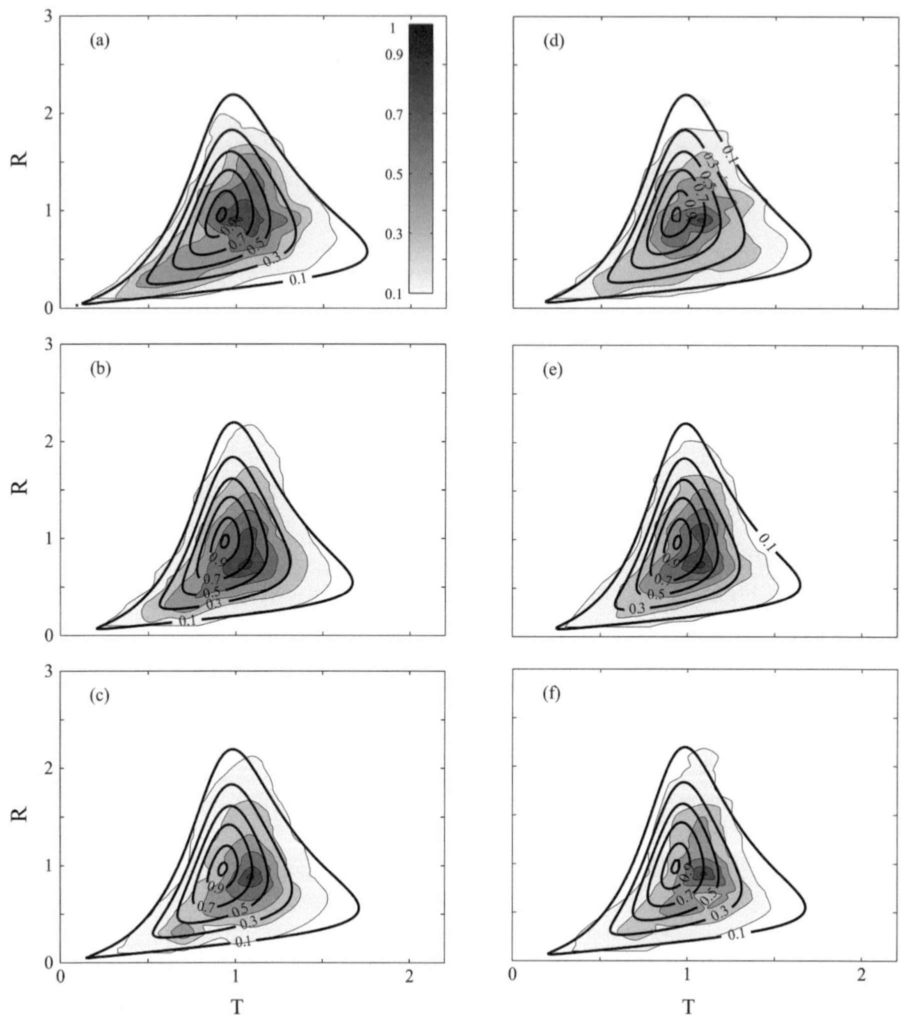

Fig. 2.6 Joint distributions of wave heights and periods at a distance of 2000 m away to the wavemaker. The three rows correspond to three cases listed in Table 2.5. On the left panel are the results in Marintek, and on the right are those in CEHIPAR

2.3.4 Exceedance Distribution of Wave Heights

As shown in Fig. 2.7, for relatively small wave heights, such as $R < 2$, both linear and nonlinear theoretical models, can describe the exceedance distribution of wave height reasonably well, and the discrepancy between them is not distinguishable. The significant deviation comes from larger wave heights where the difference between the linear

Gaussian process (solid line) distribution and the nonlinear model (dashed line) is easy to trace.

Specifically speaking, in the moderate sea states, the nonlinear MER model (Mori and Janssen 2006) predicts the larger wave heights much better than the linear model, as shown in Fig. 2.7a, b, d, e. The differences between theoretical model distributions in

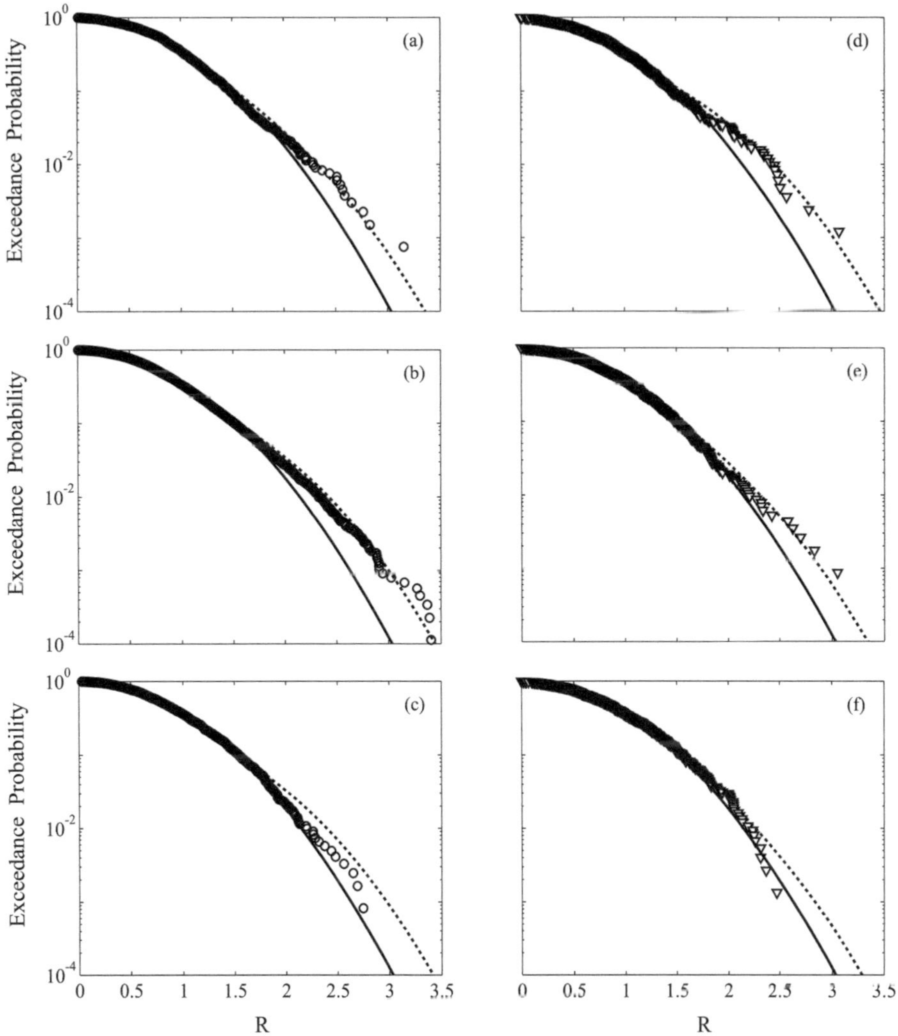

Fig. 2.7 Exceedance distributions of wave heights at a distance of 2000 m away to the wavemaker. The arrangements of the figures are the same as those in Fig. 2.6. The solid and dashed lines represent Rayleigh and MER models, respectively

the two basins are induced by the different values of the coefficient of kurtosis calculated from the measured series.

Since the coefficient of kurtosis is a statistical indicator for the high-order nonlinear effects and is related with the rogue wave density (Guedes Soares et al. 2004; Onorato et al. 2006), it is susceptible to many factors such as the initial sea state condition, evolutionary time and measured position. Anyway, to a certain degree, it can be said that the distributions of larger wave heights generated in the two wave basins are consistent in moderate sea states.

In the severe sea state, both Fig. 2.7c, f present a significant deviation from the third-order nonlinear model and get close to the exceedance distribution of the Gaussian process. This discrepancy between the nonlinear theoretical prediction and the experimental result is probably partially caused by the wave-breaking phenomenon, which dissipates wave energy and decreases the formation of more giant waves. However, it still can be argued that the time series produced in the two wave basins present a similar distribution tendency over the larger wave height.

2.3.5 Probability Distribution of Wave Periods

Comparisons of wave periods are displayed in Fig. 2.8, where a thick solid line represents the theoretical distribution (Longuet-Higgins, 1983) expressed by Eq. (2.8). Comparing the experimental results on the left panel with those on the right panel reveals that the probability distributions of wave periods in the two wave basins are compatible in the same sea state.

It is noted that the theoretical distribution corresponding to Eq. (2.8) only depends on the wave spectral width parameter ν. As mentioned, the spectral width parameter changes little among the six wave series, thus leading to an insignificant variation in the theoretical distributions of periods.

However, in both wave basins, the wave period density varies largely in different sea states, manifesting various degrees of deviation from the theoretical model, which seems consistent with the wave steepness change. This means that the theoretical model cannot describe the observed results well, especially in cases with large wave steepness, such as those presented in Fig. 2.8c, f and. In other words, the inaccurate description of wave period distribution results in the main deviation in the joint distribution of wave heights and wave periods.

Moreover, comparing Fig. 2.8(d) with Fig. 2.8(f), it is observable that the probability distribution of wave periods becomes narrower as the wave steepness increases. To some extent, the conclusions about the variation of wave period distribution are supported by the change in the wave frequency spectrum that is narrower in steeper sea states. Another conspicuous phenomenon is that all the experimental results tend to deviate to the right side compared to the theoretical predictions.

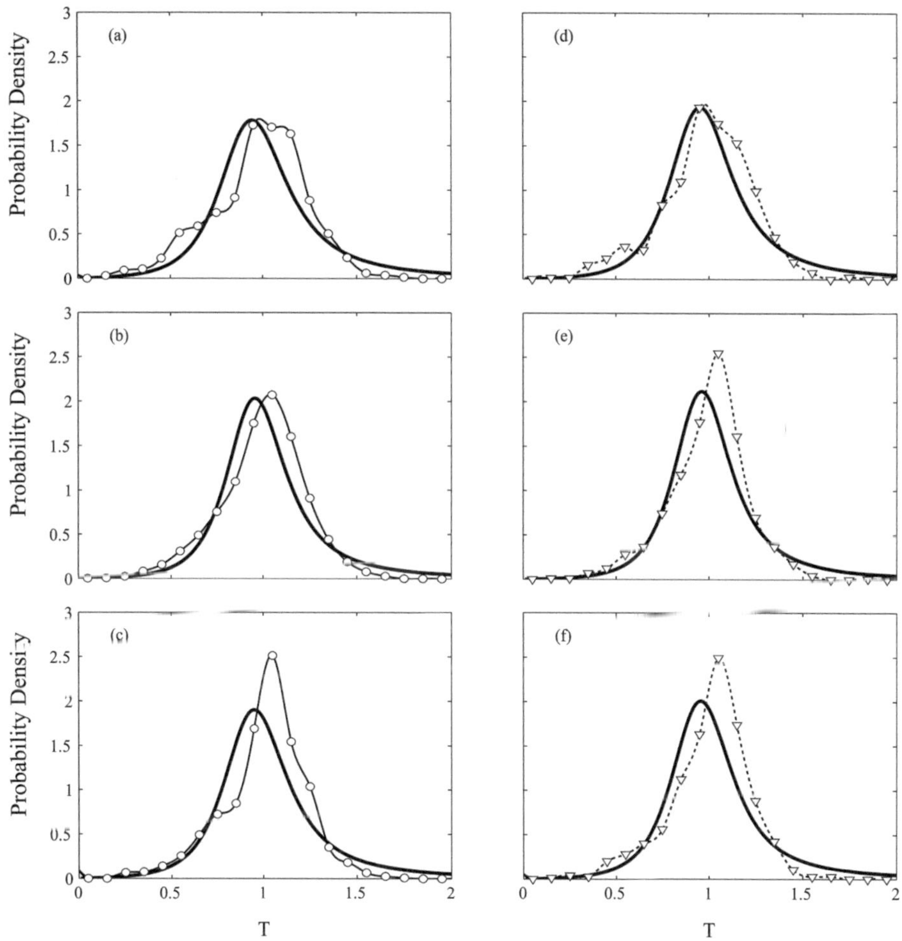

Fig. 2.8 Probability distributions of wave periods at a distance of 2000 m away to the wavemaker. The arrangements of the figures are the same as those in Fig. 2.6. The thick solid line represents the theoretical model. Thin solid and dashed lines connect the experimental data in different basins

Until now, there has not been much attention paid to the wave period study due to less influence in engineering but more complexity in distribution. It is confirmed here that the wave period distribution cannot be described well by the existing model that only involves the wave spectral width parameter, which does not vary significantly in different random sea states, and that the critical issue in improving the joint distribution is the description of wave period rather than wave height.

Wave steepness can be an alternative to wave period in the joint distribution. For narrow-band sea waves, the joint probability densities of wave steepness and wave height can be derived from the Longuet-Higgins joint probability density of wave period and wave amplitude. The bivariate probability density function can also be fitted with data by various forms of copula functions, as well as the bivariate gamma distribution function. However, the fitting quality is reasonably good only in the case of a wave system with weak nonlinearity (Antao and Guedes Soares, 2016).

2.4 Modified Theoretical Models for Joint Distributions

Motivated by the discussion in Sect. 2.3.3, two types of modified Longuet-Higgins joint distribution, presented in Sects. 2.1.3 and 2.1.4, will be further researched herein. To make a comprehensive analysis, the experimental results of cases 8201, 8202 and 8219 listed in Table 2.2 will be compared with the predictions of the modified theoretical models. The three cases can represent smooth, moderate and severe sea states and will be marked by square, circle and triangle, respectively.

2.4.1 Spatial Variation of Basic Wave Parameters

As shown in Fig. 2.9a, it is apparent that the wave spectral width decreases along the wave basin due to energy dissipation, mainly reflected in the reduced density of saturation range in the wave spectrum (Fedele et al. 2010). The discrepancy in the same location is so small that the spectral width parameter cannot be used to identify the difference among these sea states. Moreover, it is evident that $\nu_{\max} < 0.6$, i.e., all the wave series generated in the laboratory are strictly narrow-banded.

In contrast to the spectral width parameter, the disparities of steepness in different cases are so pronounced, as illustrated in Fig. 2.9b, that the difference of these sea states can be clearly identified. Furthermore, the decreasing tendency of wave steepness along the offshore basin is most strikingly evident in the severe sea state due to the serious

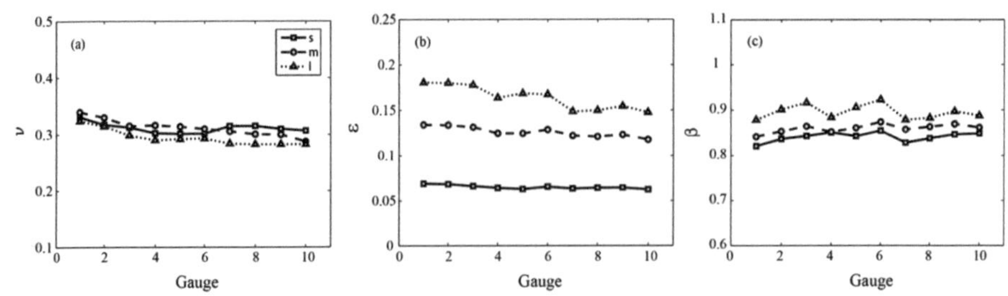

Fig. 2.9 Spatial variations of basic wave parameters. **a** wave spectral width parameter, **b** wave steepness, **c** the ratio of mean period to peak period. Square, circle and triangle represent 8201, 8202 and 8219 in Table 2.2, respectively

energy dissipation, mainly wave breaking. In contrast, the low sea state is obscured and thus almost maintains a constant level as the wave propagates downstream in the wave basin.

Figure 2.9c strongly supports the conclusion that the different representative wave periods are interrelated, particularly for the mean and peak periods, considering that a similar approximately linear regression model exists in all sea states. Both variables are typical such that the issue of which one being used to work as a parameter in describing the sea state has already been argued for a long time.

The wave periods will usually decrease along the wave basin (Zhang et al. 2013). As discussed before, The slightly increasing tendencies in all sea states mean that the mean period approaches the peak period downstream in the wave basin due to the decreased spectral width. Furthermore, comparing the values of β the same location among different cases reveals that as the sea state becomes more and more severe, the mean wave period will gradually approach the peak period as well.

2.4.2 Comparison of Joint Distributions

To be concise, in the following sections, the comparison will focus only on the measurements at gauge 10 where the nonlinear effect has been fully developed in all sea states.

Not only the original Longuet-Higgins (1983) joint distributions, expressed in Eq. (2.4), but also the other two modified theoretical joint distributions represented by Eqs. (2.12) and (2.20) are compared with the observed statistics in Fig. 2.10 and Fig. 2.11, respectively. Here, contours take the value (0.99, 0.9, 0.8, 0.6, 0.4, 0.2) $\times p_t$ (or p_o) from the centre contour outwards for both theoretical distribution (heavy solid line) and observed result (coloured area) where p_t (or p_o) has the same meaning as before, i.e., the theoretical (or observed) value of density mode.

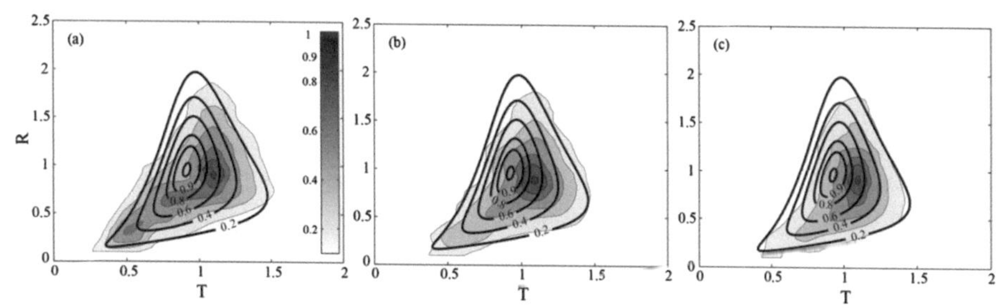

Fig. 2.10 Original Longuet-Higgins joint distributions versus experimental results at Gauge 10 in three typical sea states. **a**, **b** and **c** correspond to cases 8201, 8202 and 8219, respectively

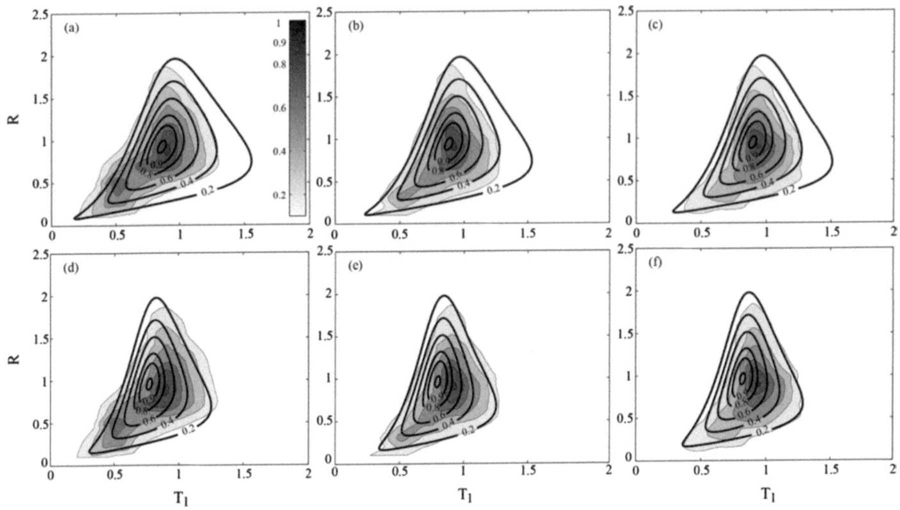

Fig. 2.11 Modified Longuet-Higgins joint distributions versus experimental results at Gauge 10 in three typical sea states. The three sequent columns correspond to cases 8201, 8202 and 8219, and the theoretical distributions (solid contour line) in the two rows are from Eqs. (2.12) and (2.20), respectively

Figure 2.10 reveals that the original Longuet-Higgins joint distributions deviate from the experimental results in all sea states, especially around the density mode. Moreover, the theoretical joint distributions do not display too much difference among different sea states because they are determined only by the wave spectral width parameters, which are almost equal to each other, as indicated clearly in Fig. 2.9a.

However, the observed statistics of joint distributions present a large discrepancy among these three sea states, implying that the Longuet-Higgins joint distribution model should be further improved. Up to now, few efforts have been made on this subject, and little improvement has been achieved due to the rapidly increasing complexity of simultaneously involving two correlated random variables in one distribution function.

Apparently, in the upper panel of Fig. 2.11, i.e., (a)–(c), the modified Longuet-Higgins joint distribution I expressed by Eq. (2.12) is able better to catch the location of density mode under all circumstances even though the area of the contour is enlarged for the lower probability density domain. Thus, it is reasonable to believe that this simplified theoretical model can predict the most frequent combination of wave period and wave height in a time wave series.

Meanwhile, in the lower panel of Fig. 2.11, i.e., (d)–(f), it can be concluded that the modified Longuet-Higgins joint distribution II represented by Eq. (2.20) is an alternative theoretical model of the original Longuet-Higgins joint distribution for the reason that the deviations of the location of density mode are slightly decreased but still significant.

Anyway, the alternative form of Longuet-Higgins joint distribution, which can be potentially applied in mixed sea states (Rodriguez and Guedes Soares 2001), has been correctly formulated herein.

Moreover, it seems impossible to eliminate the difference between theoretical model predictions and laboratory observations in the linear wave theory. In other words, the nonlinearity of the wave series mainly leads to these disagreements, which are clearly indicated in the domain around the density mode.

On the other hand, the conclusions drawn above are supported by all recently derived theoretical models (e.g., Stansell et al. 2004) for the reason that they are all based on the same linear assumptions as those used in Longuet-Higgins joint distribution, but there never exists a perfect model applicable to all sea states or in other words, those improvements are to a certain degree confined to some specified aspects. To propose a better joint distribution, the nonlinear theory has to be introduced, especially for the distribution of wave period.

2.5 Dynamic Pressure Under Crests of Nonlinear Waves

2.5.1 Introduction

For the safe design and operation of ships, it is critical to precisely predict the hydrodynamic loads induced by various ocean waves. One of the load components, the Froude-Krylov force, is directly caused by the incident wave and computed by integrating the dynamic pressure over the wetted surface of the floating body. This has been determined as being the main nonlinear component of several nonlinear time domain codes that assess wave-induced loads. Therefore, the exact information on wave dynamic pressure, especially under the large wave crest, is essential for accurately estimating wave forces.

For regular waves with infinitesimal amplitude, the standard wave theories such as Airy theory or Stokes wave theory are assumed to be valid only up to the still water level, but practically, they can often work outside of the range of validity. Specifically speaking, if the nonlinearity is very weak, the dynamic pressure below the mean water level can be predicted accurately by these theories, and the pressure above the mean water level can also be reasonably estimated using the Taylor series. However, for the finite or even large amplitude wave, the free surface pressure boundary condition can no longer be satisfied in these simplified wave theories. Thus, the theoretical prediction error will be increased to various degrees.

To solve this problem, a wide range of correction terms have been proposed to modify the form of the transfer functions in these theoretical expressions, which can improve the accuracy of pressure predictions (Tsai et al. 2005). The most common technique used is the stretching method, for instance, the Wheeler-Stretching, the Chakrabarti-Stretching, the Delta-Stretching methods, and the Linear-Extrapolation approach. A new adaptive second-order stretching approach has been proposed to modify the dynamic pressure distribution above the mean water level, showing a better performance than the previous models (Clauss et al. 2009).

However, the above empirical methods are insufficient to describe the pressure distribution under the surface elevation more accurately; thus, the nonlinear wave theories are indispensable. One complex function for the Stokes wave was first derived by Escher and Schlurmann (2008), which cast some light on the impact of the nonlinear effect on the dynamic pressure. Using the elementary harmonic function theory, the variation trend of pressure beneath the Stokes wave can be further analysed (Constantin and Strauss 2010). Based on the implicit function theorem, more complex equations can be obtained to relate the pressure at the bottom of the fluid to the surface elevation of a travelling wave without any approximation (Kogelbauer 2015; Oliveras et al. 2012). Nevertheless, even though it is numerically possible, solving these equations is computationally expensive.

Moreover, the behaviour of the pressure beneath a solitary wave has also been investigated, and a good agreement was attained between the theoretical predictions and experimental measurements (Constantin et al. 2011). However, the pressure presented is the total pressure rather than the dynamic one, of which the variation trend cannot be identified due to the influence of the predominant hydrostatic pressure. Besides, with a conformal hodograph transform, an explicit formula can be theoretically derived to relate the profile of a solitary water wave with pressure data measured at the flatbed of the fluid domain, valid for both small and large amplitude waves.

Until now, the theoretical analysis of dynamic pressure has only been limited to regular waves or solitary waves, and the analytical expression has become very complex after considering the fully nonlinear effect. Extending a similar analysis to irregular waves is very difficult in theory. Therefore, numerical simulations or laboratory experiments seem to be more feasible for exploring the subtle and complex dynamic pressure in irregular sea states.

Usually, in the context of potential theory, the numerical simulation solves the problem of wave propagation on the free surface. Consequently, the information on the pressure field is not directly available. Hence, specific procedures such as the Dirichlet to Neumann Operator (Bateman et al. 2012) or direct inversion of a matrix can be incorporated into the wave model to derive the corresponding pressure beneath the surface wave. One successful example, shown in previous work, is reconstructing the vertical profiles of dynamic pressure under a freak wave crest simulated by different numerical wave models (Ducrozet et al. 2016b).

Within the framework of Serre-Green-Naghdi equations and fully nonlinear potential equations, the pressure distribution beneath transient wave packets is also investigated (Touboul and Pelinovsky 2018), and the roles of nonlinearity and nonlinear dispersion in reconstructing the relationship between pressure and surface elevation has been discussed. By solving the 2D Navier–Stokes equation (Hu et al. 2021), the dynamic pressure in

Stokes and high-order rogue waves shows a similar distribution but different extreme values under the maximum crests and minimum troughs.

As to the laboratory experiment, the first rough comparison of dynamic pressure was made between the New Year wave (Cherneva and Guedes Soares 2008) and the Peregrine breather solution (Klein et al. 2016). The variation trend is very similar, but the discrepancy increases with the water depth, which is mainly attributed to the different periods of these two giant waves. Moreover, the comparison is only limited to the region below the mean water level without considering the critical crest region above the mean water level.

Note that in numerical simulations and laboratory experiments, a similar variation trend of dynamic pressure is detected below the mean water level, i.e. exponentially decayed with increased depth (Constantin 2016), while above the mean water level, the opposite change tendency can be widely observed in different papers, i.e. decreasing to zero (Clauss et al. 2009) versus increasing to maximum (Ducrozet et al. 2016b) at the free surface, mainly due to the controversial definition of dynamic pressure in this region (Zhang et al. 2023).

2.5.2 Theoretical Pressure Models for Regular and Irregular Waves

The pressure p at any point in the wave field can be obtained from the linearized Bernoulli equation:

$$p = -\rho g z - \rho \frac{\partial \phi}{\partial t}. \tag{2.26}$$

where ρ is the water density, and z is the location below the mean water level ($z \leq 0$). Substituting the velocity potential function ϕ of small amplitude waves into Eq. (2.26) and neglecting the velocity-squared terms, the following expression can be derived:

$$p = -\rho g z + \rho g \frac{H}{2} \frac{\cosh k(z+h)}{\cosh kh} \cos(kx - \omega t). \tag{2.27}$$

where h is the absolute water depth, and H denotes crest-to-trough wave height. It can be seen that the total pressure p consists of two parts. The first term on the right-hand side of the Eq. (2.27) is the hydrostatic pressure, and the second is the dynamic pressure.

Under the wave crest, the formulation of dynamic pressure is only valid in the domain from the bottom up to the mean water level, and its maximum value appears at $z = 0$, equal to $\rho g C$ with C being the crest height. Note that $C = H/2$ in linear wave theory. To determine the pressure above the mean water level, a Taylor series for a small positive distance must be applied to Eq. (2.27) and to the first order, the following formulation is derived:

$$p = -\rho g z + \rho g \eta. \tag{2.28}$$

Thus, to this approximation, the pressure is hydrostatic under the wave crest down to $z = 0$. Below that depth, however, it deviates from the hydrostatic law.

There are two different definitions for the dynamic pressure in Eq. (2.28). One type, widely adopted in numerical simulations, is to define $\rho g\eta$ as the dynamic pressure and thus the term$-\rho gz$ is the hydrostatic pressure which is negative above the mean water level. The other type is to consider both terms in Eq. (2.28) as the dynamic pressure, which is more consistent with the physical reality and commonly used in laboratory experiments. That is why the different dynamic pressure distributions can be widely observed in many works in the literature. In this subsection, the second type of definition is used to measure the dynamic pressure in all tests.

In irregular sea states, the corresponding dynamic pressure is simply synthesized by the linear superposition of the pressure associated with each harmonic wave component, which can be obtained by utilizing the Fourier transform technique:

$$p = -\rho gz + \sum_{i=1}^{N} \rho g a_i \frac{\cosh k_i(z+h)}{\cosh k_i h} \cos(k_i x - \omega_i t). \tag{2.29}$$

2.5.3 Layout of the Experimental Setup

The wave flume experiments were conducted at the Shandong Provincial Key Laboratory of Ocean Engineering at the Ocean University of China. The flume size is 60.0 m long, 3.0 m wide and 1.5 m high, respectively. The overall arrangement in the wave flume is schematically shown in Fig. 2.12, and the mid-position of the pressure measurement zone is located 26.3 m away from the wavemaker. Note that the water depth is fixed at 1.0 m, and no significant transverse variation can be detected for all cases.

At one end of the wave flume, the reciprocating motions of a piston-type wavemaker are driven by a low-inertia servo motor and controlled by a computer to generate the user-defined waveforms. Programming the computer-based control system can create the signal for generating linear and nonlinear focused waves. A triangular-shaped wave absorber, made of porous media, is installed at the opposite end to absorb the incident waves passively.

It is well known that reflection could be a reason for the non-homogeneous conditions along the wave flume (Olfateh et al. 2017). In this subsection, the reflection coefficient has been evaluated firstly with different methods such as Mansard three-point method, Isaacson three-point method, Transfer Function method and Hilbert method, and the maximum reflection coefficient is not larger than 5% in all tests, thus excluding the influence of reflected waves.

Along the wave propagation direction, 15 capacitive wave gauges are deployed to record the surface elevations, 10 of which are densely arranged around the location of the

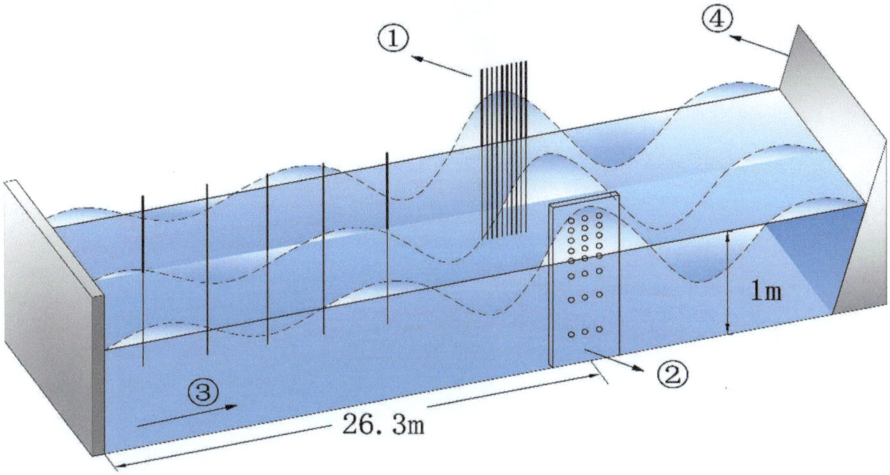

Fig. 2.12 Sketch of the 3D wave flume: ① wave gauges, ② pressure measuring device, ③ incoming wave direction, ④ sloping beach

pressure measuring device, mainly for accurately catching the profile of focused waves and largely reducing the time required to adjust the focal location and maximum crest height. To measure the dynamic pressure more precisely, 30 pressure transducers are deployed at this location and fixed on acrylic plates, as shown in Fig. 2.13. In the horizontal direction, the pressure transducers are arranged in three columns with an identical interval of 5 cm. In contrast, they are not uniformly distributed in the vertical direction, with more transducers being located around the mean water level to capture the maximum pressure under the wave crest as accurately as possible.

The local physical layout for the wavefield measurement is illustrated in Fig. 2.13b. Both wave gauges and pressure transducers are placed close to the side wall of the flume to avoid disturbing the pressure field in the measuring zone. The cylindrical probes are precisely mounted into the circular holes in the acrylic plate to form a smooth face without affecting the wave field (left-hand side of the plate in Fig. 2.13b). Thus, the pressure in the measuring zone can be recorded by the top surface of these probes based on the assumption that the dynamic pressure at the same vertical location is identical in the transverse direction of wave propagation.

One wave gauge (i.e., No. 10) is purposely aligned with the middle column of pressure transducers to record the corresponding surface elevation. Even though the acrylic plate is very thin, the oscillating motion induced by the incident wave is still very significant. To solve this problem, a fixing device is installed on the top of the plate.

The sampling frequency of the wave gauge and pressure transducer is set to 50 Hz in all tests, and the duration of each realization, including ramping time, is at most 150 s for both regular and irregular waves to avoid the potential influence of reflected waves. For

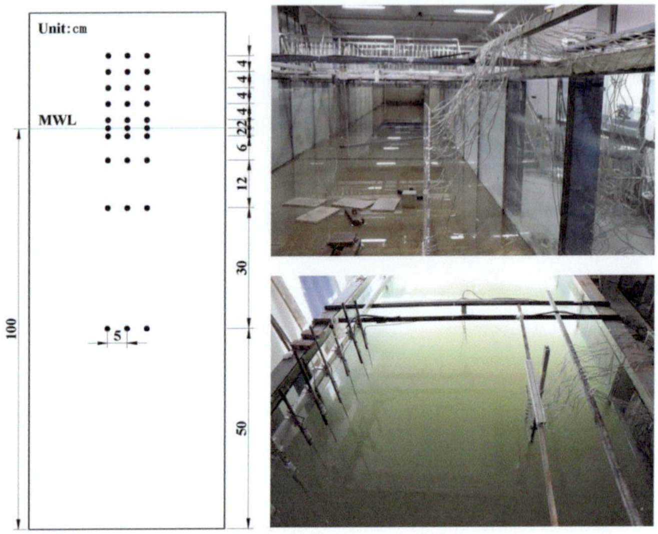

(a) Dimensions of device (b) Layout of field measurement

Fig. 2.13 Detailed arrangement for pressure measurement

each case, the same test is repeated three times to ensure the repeatability and accuracy of the data.

2.5.4 Dynamic Pressure in Regular Waves

The dynamic pressure in regular waves is first investigated, and the corresponding wave parameters in the experiments are listed in Table 2.8. Note that the wave steepness in this study is defined as the product of crest height and wave number.

The first 10 test cases (R1 ~ R10) have the same crest height but different wave periods, and thus the influence of wave period or wavelength on the dynamic pressure can be studied. The last two cases (R11 and R12) and the first one (R1) can be compared to show the effect of crest height on the dynamic pressure at the identical wavelength. The

Table 2.8 Parameters of regular waves

Case	R1	R2	R3	R4	R5	R6	R7	R8	R9	R10	R11	R12
Crest (cm)	6.5	6.5	6.5	6.5	6.5	6.5	6.5	6.5	6.5	6.5	4.9	5.5
Trough (cm)	5.0	5.1	5.2	5.3	5.3	5.5	5.6	5.6	5.8	5.8	4.0	4.3
Period (s)	0.9	1	1.1	1.2	1.3	1.4	1.5	1.6	1.7	1.8	0.9	0.9
Steepness	0.30	0.26	0.21	0.18	0.16	0.14	0.12	0.10	0.09	0.09	0.24	0.27

above two types of comparisons (fixed crest height or fixed period) can also reveal the influence of nonlinearity on the dynamic pressure as indicated by the wave steepness given in Table 2.8. It has to be mentioned that P_s denotes the theoretical maximum dynamic pressure predicted by the linear wave theory, which is only dependent on the crest height.

To comprehensively demonstrate the variation of dynamic pressure along a whole wavelength at different vertical locations, one particular example is first presented in Fig. 2.14, where the experimental results are derived with a phase average technique over the wave-coherent component after performing a triple decomposition of the pressure signal (Buckley and Veron, 2016). The significant discrepancy from the linear theoretical prediction (LTY) is mainly observed under the wave crest and trough. Therefore, to be concise in the following comparison, only the vertical profiles of dynamic pressure at these two instants are discussed.

Figure 2.15a shows the measured maximum dynamic pressure extracted from a set of regular waves (R1 ~ R10) with identical crest height. The maximum pressure increases with the wave period, but the growth rate gradually decreases.

To be specific, the first three cases denoted by the empty circles represent the deep-water waves. As the wave nonlinearity decreases (i.e., increasing period), the maximum dynamic pressure rises quickly. However, a similar variation trend can still be detected for the intermediate water waves marked by the full circles, but it is not as evident as the deepwater waves. Hence, it can be concluded that the relative water depth also affects the maximum dynamic pressure. Moreover, due to the nonlinear effect, the maximum

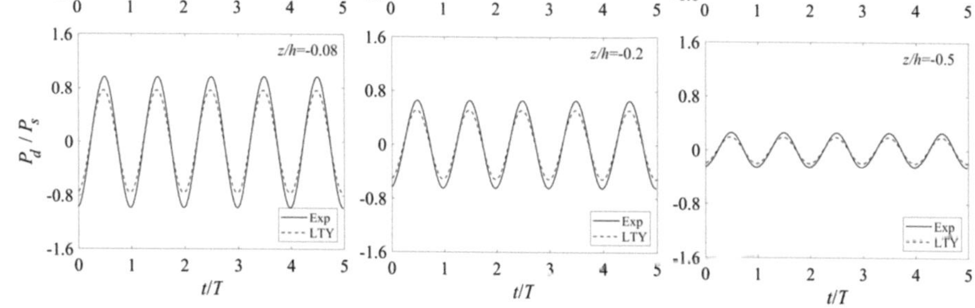

Fig. 2.14 Comparison of dynamic pressures P_d between laboratory observation and linear theoretical prediction at different vertical locations in the case of R4. P_s denotes the maximum dynamic pressure predicted by the linear wave theory. T is the wave period

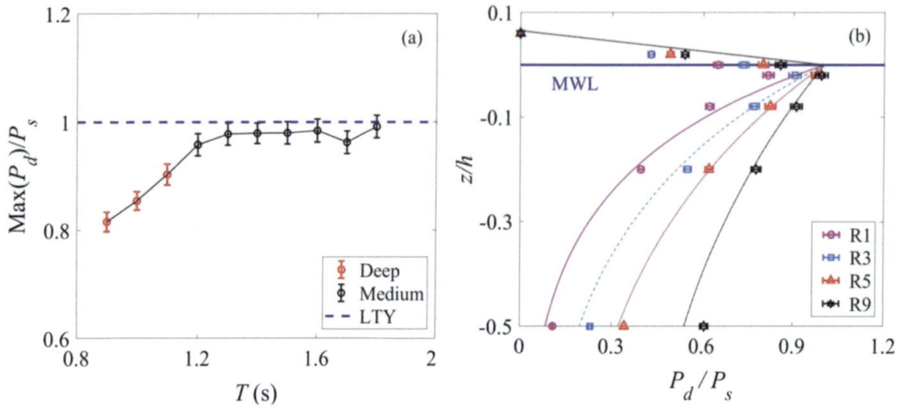

Fig. 2.15 Dynamic pressure P_d under the crest of a regular wave with the same crest height but different wave periods T: **a** maximum pressure, **b** vertical pressure profile. Note that the curves in (b) depict the corresponding pressure predicted by the linear wave theory. P_s has the same meaning as that in Fig. 2.14

dynamic pressures in all cases are smaller than the corresponding linear theoretical value (LTY), equal to one irrespective of wave period and water depth according to Eq. (2.28). This deviation is consistent with previous numerical simulation results of the fifth-order Stokes wave (Hu et al., 2021).

Furthermore, four typical pressure profiles under wave crests are illustrated in Fig. 2.15b. The corresponding curves also compare and depict the theoretical values predicted by the linear wave theory. The locations of maximum dynamic pressure are slightly lower than the theoretical ones, i.e. the man water level (MWL) in all cases. Above the mean water level, the linear wave theory always overestimates the dynamic pressure. Below the mean water level, the linear wave theory tends to underestimate the dynamic pressure to various degrees, depending on the wave nonlinearity.

Taking the case R1 as an example, if the nonlinearity is very strong, the starting depth of underestimation of dynamic pressure is far below the mean water level. Thus, the ending depth of overestimation is accordingly extended. Finally, it can be concluded that even though the crest heights of regular waves are identical, the dynamic pressure field under the crests can be vastly different, mainly due to the effect of wave nonlinearity.

The corresponding negative dynamic pressures under wave troughs are presented in Fig. 2.16. The variation trend of the absolute value of minimum negative pressure, depicted in Fig. 2.16a, is very similar to that detected in the maximum positive pressure presented in Fig. 2.15a. Nevertheless, the difference can also be observed, mainly reflected in the magnitude that seems more significant in the negative pressure in all cases.

As before, four typical pressure profiles under wave trough are further illustrated in Fig. 2.16b. The linear wave theory represented by the curve underestimates the magnitude

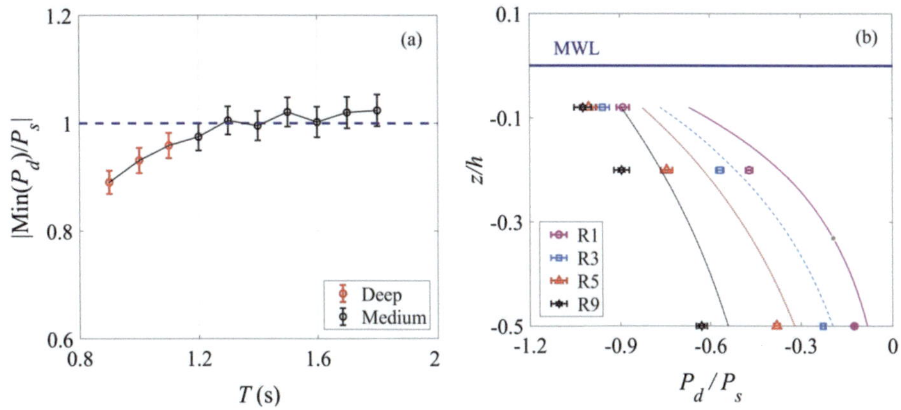

Fig. 2.16 Dynamic pressure P_d under the trough of a regular wave with the same crest height but different wave period T: **a** minimum pressure, **b** vertical pressure profile. P_s and the smooth curves have the same meanings as those in Fig. 2.15

of negative dynamic pressure in all cases. The most significant discrepancy appears in the case of R1 at the location nearest to the free surface due to the most substantial nonlinearity. In agreement with the discrepancy observed in positive dynamic pressure under wave crests, the difference of negative pressure between theoretical prediction and experimental measurement is also quickly reduced with the increased depth, especially pronounced in case R1, where the discrepancy is changed from 22.4% to 10.5% at the same vertical distance. Thus, it can be inferred that the influence of nonlinear bound wave components on the dynamic pressure is mainly limited to the region around the mean water level.

To further explore the effect of nonlinearity on the dynamic pressure without the influence of relative water depth, three typical cases with the same wave period are considered in this subsection, and the pressure profiles under wave crest and wave trough are compared in Fig. 2.17, respectively. Note that the measured pressures have been normalized to exclude the discrepancy induced by wave amplitude difference. That is, $\overline{P}_d = P_d/B$ and $P_s = \rho g$ where B is the corresponding crest height or trough depth.

Overall, with increased nonlinearity, the positive dynamic pressure per unit height is gradually decreased under wave crests (Fig. 2.17a) while the magnitude of negative dynamic pressure per unit depth is mildly increased under wave troughs (Fig. 2.17b). In other words, the growth rate of pressure magnitude with respect to wave amplitude is different under wave crest and wave trough. In this sense, it can be argued that dynamic pressure variation under wave troughs is more sensitive to wave nonlinearity than under the wave crests. This conclusion can also be confirmed by the discrepancy between the linear theoretical predictions (LTY) and the laboratory measurements (symbols), which is

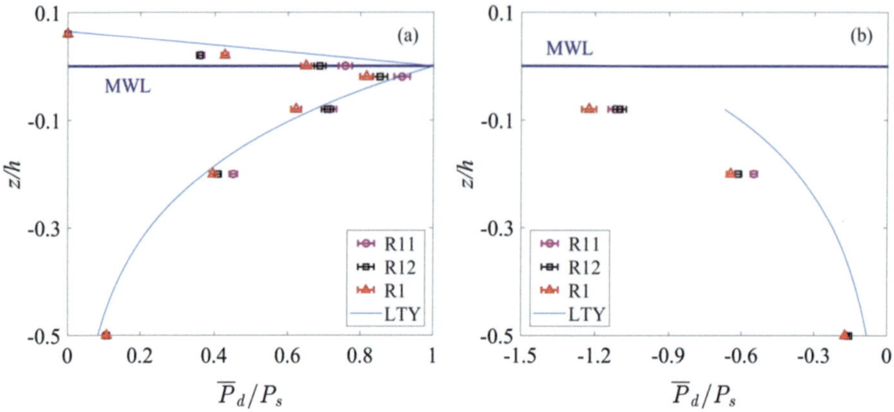

Fig. 2.17 Profiles of normalized dynamic pressure under the crest (**a**) and the trough (**b**) of a regular wave with the same wave period. $\overline{P}_d = P_d/B$ and $P_s = \rho g$ where B is the corresponding crest height or trough depth

larger under the wave troughs than under the wave crests, up to a maximum difference of 57.1%.

In summary, nonlinearity has played a different role in modifying the dynamic pressure under different parts of regular waves. Under wave troughs, it increases the amplitude of negative pressure. Under wave crests, it decreases the positive pressure in the upper section but increases the pressure in the lower section with the boundary determined by the nonlinearity strength. Generally speaking, for the case with strong nonlinearity, the upper section where the pressure decreases tends to be very large.

2.5.5 Dynamic Pressure in Irregular Waves

To be more consistent with reality, the dynamic pressure of individual waves with crest height equal to 6.5 cm is further investigated in irregular sea states. The critical parameters are listed in Table 2.9, and the sea state is characterized by the JONSWAP spectrum with the peak enhancement parameter $\gamma = 3$ in all five cases.

Table 2.9 Parameters of irregular sea states

Case	IR1	IR2	IR3	IR4	IR5
H_s (cm)	12	12	12	12	12
T_p (s)	1.0	1.2	1.4	1.6	1.8
$H_sk_p/2$	0.24	0.17	0.13	0.10	0.08

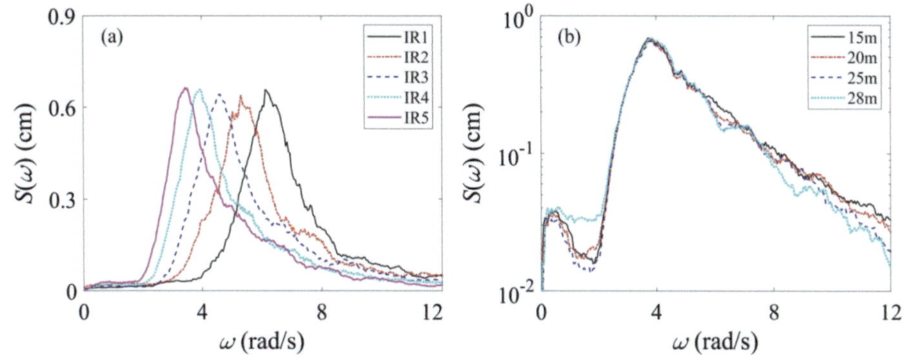

Fig. 2.18 **a** Amplitude spectra of the time series for wave analysis and **b** spatial evolution of the wave spectra in the wave flume

The smoothed amplitude spectra of time series for wave analysis in five irregular sea states have been shown in Fig. 2.18a. The desired spectral peak frequencies are exactly reproduced in the laboratory experiment, and the significant wave heights, calculated with the zeroth-order spectral moment, are almost the same as those listed in Table 2.9. The spatial evolution of frequency spectra in Fig. 2.18b reveals that the high-frequency components are gradually reduced along the wave channel due to energy dissipation due to friction. However, the peak frequency domain is almost identical at different locations, which can ensure the homogeneity of the wave field along the wave channel.

Considering the variability and uncertainty in irregular sea states (Bitner-Gregersen et al. 2021; Toffoli et al. 2017), all dynamic pressures under wave crests with a height of 6.5 cm are extracted from the measured data to construct a sample for statistical analysis (Bitner-Gregersen et al. 2022). The fitted relationship between the maximum pressure and the trough-to-trough period is shown in Fig. 2.19a. Overall, the maximum dynamic pressure under the wave crest still increases with the wave period, in agreement with the previous conclusions drawn from regular waves shown in Fig. 2.15a. However, in most cases, the pressure magnitude in irregular waves tends to be larger than that presented in regular waves and even exceeds the linear theoretical prediction (LTY). Besides, the maximum pressure seems to fluctuate greatly at identical wave periods.

To explain the above phenomenon in Fig. 2.19a, the following correlation coefficients R_η and R_{Pd} are introduced in Fig. 2.19b to evaluate the relationship between the waveform and the dynamic pressure.

$$R_\eta(\eta_1, \eta_2) = \frac{C(\eta_1, \eta_2)}{\sqrt{C(\eta_1, \eta_1)C(\eta_2, \eta_2)}}, \tag{2.30}$$

$$R_{Pd}(P_1, P_2) = \frac{C(P_1, P_2)}{\sqrt{C(P_1, P_1)C(P_2, P_2)}}. \tag{2.31}$$

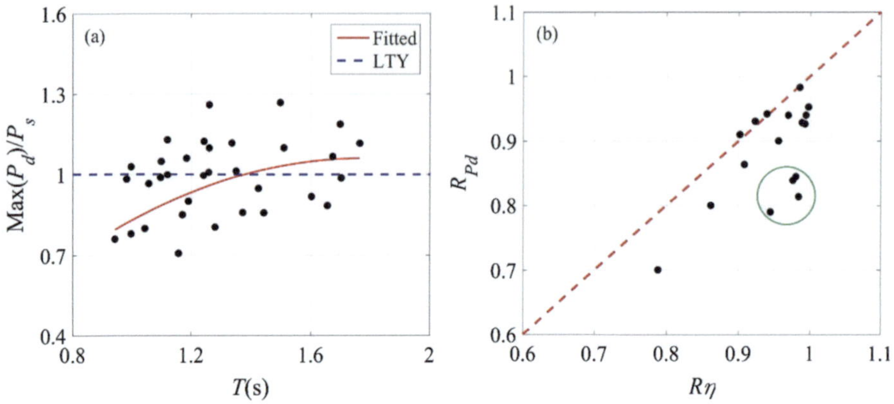

Fig. 2.19 **a** Relationship between the trough-to-trough period T and the maximum pressure P_d under the wave crest equal to 6.5 cm, **b** relationship between two correlation coefficients (dynamic pressure vs. surface elevation) for individual waves with almost the same crest height and trough-to-trough period. P_s has the same meaning as that in Fig. 2.14

The correlation coefficient of dynamic pressure R_{Pd} is generally proportional to the correlation coefficient of waveform R_η, indicating that the variation of dynamic pressure under wave crest is closely related to the change of individual wave profile. However, some data points in Fig. 2.19b fall below the red dashed line, i.e. $R_\eta > R_{Pd}$, particularly pronounced for those confined to the green circle. Hence, it can be inferred that even though the profiles of two individual waves are pretty close, the corresponding dynamic pressures under the crest can differ in some cases. Therefore, in an irregular sea state, evaluating the magnitude of dynamic pressure is inaccurate just by the profile of individual waves.

To elucidate the mechanism involved in the above phenomenon, two typical examples are presented in Figs. 2.20 and 2.21, where the surface elevations of individual waves and the corresponding dynamic pressures under wave crests are compared simultaneously. Note that in each figure, two waves (shaded area) are from different sea states, but they still have identical crest height (6.5 cm) and trough-to-trough periods. Although above the mean water level, the surface elevations of individual waves are almost the same, the dynamic pressures under wave crests (vertical dashed line) are substantially different, mainly due to the influence of the rear (Fig. 2.20) and front (Fig. 2.21) deep troughs, which also belong to the surrounding individual waves with different characteristic parameters.

Therefore, to take into account the influence of all individual waves within the same wave group on the dynamic pressure, the statistical parameter such as peak wave period, rather than the individual trough-to-trough period, is more suitable for the estimation of dynamic pressure under wave crest in irregular sea state.

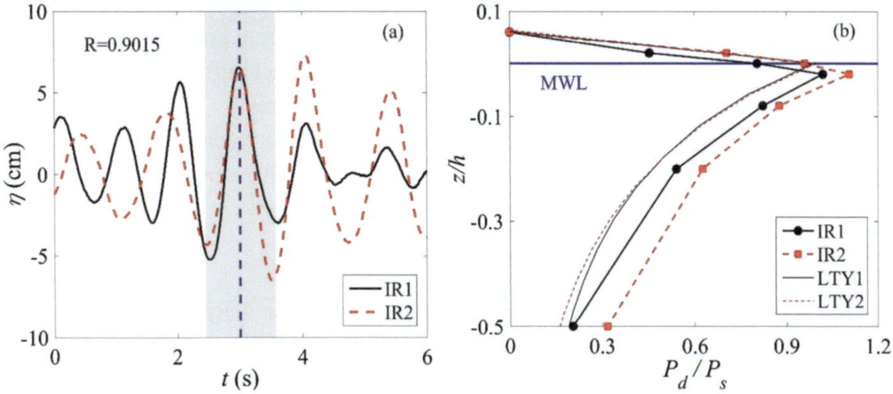

Fig. 2.20 Comparison on the profiles (**a**) of individual waves (shaded area), which have the same crest height (6.5 cm) and trough-to-trough period (1.12 s), from case IR1 and IR2, and of the corresponding dynamic pressure (b) under wave crests (vertical dashed line). 'LTY1' and 'LTY2' denote the linear theoretical predictions. In the following figures, P_d and P_s have the same meanings as those in Fig. 2.14. R means the correlation coefficient of wave profiles in the shadow zone

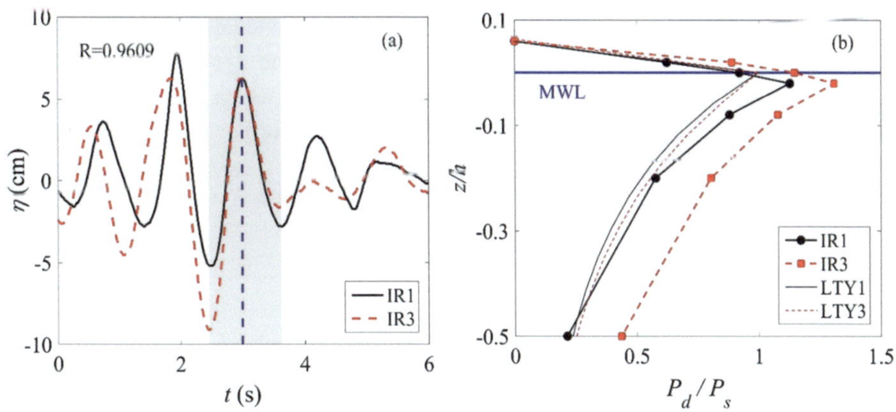

Fig. 2.21 Comparison on the profiles of individual waves (shaded area), which have the same crest height (6.5 cm) and trough-to-trough period (1.26 s), from case IR1 and IR3, and of the corresponding dynamic pressure under the wave crests (vertical dashed line). 'LTY1 and 'LTY3' denote the linear theoretical predictions, respectively

In combination with the parameters listed in Table 2.9, it can be concluded that if similar individual waves (i.e. identical crest height and trough-to-trough period) are generated in different irregular sea states, the dynamic pressures under wave crests tend to be more significant in the case with longer wave peak period.

In addition, it can be detected that the location of maximum dynamic pressure is still below the mean water level in irregular sea states. The theoretical predictions on pressure, obtained by the linear superposition method, do not show much difference between these two individual waves, although calculated in different sea states.

However, they largely deviate from the experimental measurements, indicating that the dynamic pressure under wave crest is more complex in irregular sea states and cannot be simply analysed by the first-order wave theory.

2.5.6 Comparison of Pressures Between Regular and Irregular Waves

It makes sense to compare the dynamic pressure under wave crests between regular and irregular individual waves, which have the same crest height (6.5 cm) and trough-to-trough period. Note that to find the comparable irregular and regular individual waves easily, the cases considered should have similar critical wave parameters, e.g. the spectral peak period being close to the regular wave period. To be concise, only two typical cases are selected and demonstrated in Figs. 2.22 and 2.23, respectively.

Apparently, except for the front and rear troughs, most variations of the surface elevations are consistent between the regular and irregular individual waves, quantitatively indicated by the high correlation coefficient R. However, the dynamic pressure at any location beneath the crest is larger in the irregular wave, which can be attributed to the complex interaction of individual waves during the dispersive process. The coupled

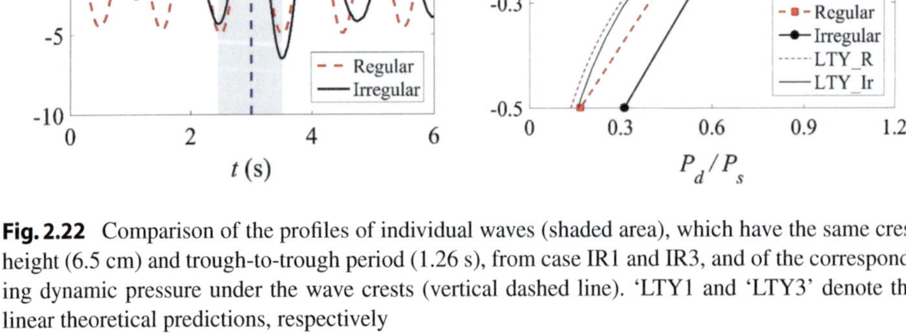

Fig. 2.22 Comparison of the profiles of individual waves (shaded area), which have the same crest height (6.5 cm) and trough-to-trough period (1.26 s), from case IR1 and IR3, and of the corresponding dynamic pressure under the wave crests (vertical dashed line). 'LTY1 and 'LTY3' denote the linear theoretical predictions, respectively

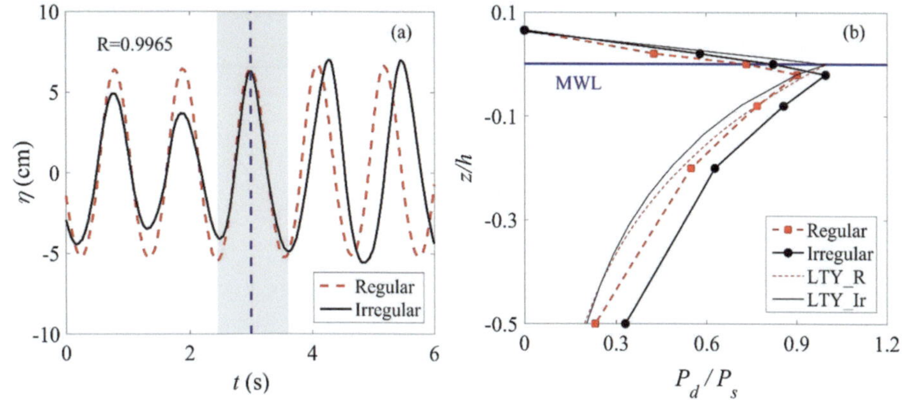

Fig. 2.23 Comparison between regular and irregular individual waves with crest height of 6.5 cm and period of 1.1 s (shaded area): **a** wave profile, **b** dynamic pressure under wave crest (vertical dashed line). 'LTY_R' and 'LTY_Ir' represent the linear theoretical predictions for regular and irregular waves

effects of various wave components can enlarge the vertical acceleration of water particles in most cases and thus lead to a larger dynamic pressure in the irregular wave field. Therefore, the theoretical predictions computed with the linear superposition method will deviate significantly from the experimental measurements compared to the regular wave discrepancy.

Hence, it can be argued that in most cases, the dynamic pressure in irregular waves tends to be larger than that in regular waves, further confirming the previous conclusion drawn from the comparison of the maximum dynamic pressures presented in Figs. 2.19a and 2.15(a), respectively. However, in some rare cases, the dynamic pressure in irregular individual waves can be smaller than that in regular individual waves. As shown in Fig. 2.24, the maximum crest height achieved by an irregular individual wave is equal to that of a regular wave, and it seems to be realized by absorbing wave energy from the surrounding individual waves within the same wave group, i.e. wave focusing.

Nevertheless, the corresponding dynamic pressure under the irregular wave crest is far smaller than that under the regular wave crest. Besides, the deviation pattern from theoretical prediction significantly differs from those detected in most irregular waves. Thus, exploring the property of dynamic pressure under the crest of focused waves is essential for irregular waves.

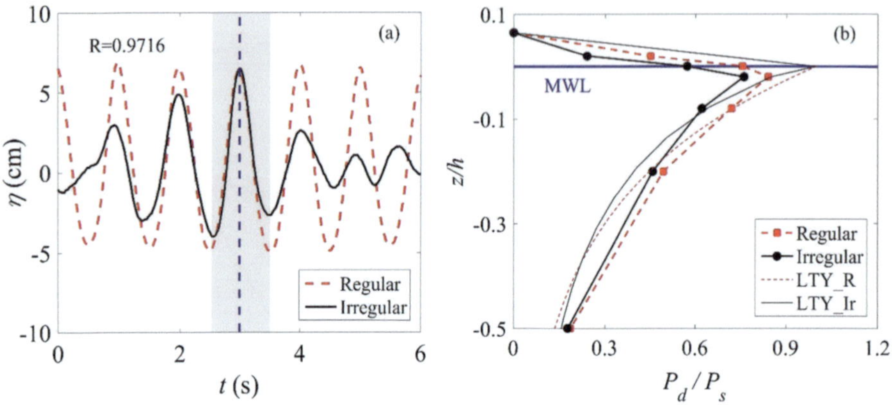

Fig. 2.24 One specific case for comparing regular and irregular individual waves with a crest height of 6.5 cm and period of 1 s (shaded area)

2.5.7 Dynamic Pressure in Focused Waves

For a better understanding of the influence of nonlinearity on the dynamic pressure in focused waves, four typical waves generated with two different focusing mechanisms are compared in this section, and the detailed parameters are given in Table 2.10 where 'LFW' and 'NFW' denote the linear and nonlinear focused waves, respectively. According to the relative water depth ($kh > 1.36$), it can be inferred that modulational instability can develop in all nonlinear focused waves.

It must be stressed that the linear focused wave is generated by the phase coherence method, and the nonlinearity involved is caused by the bound wave components (Baldock et al. 1996). The Peregrine breather solution forms the nonlinear focused wave, and the modulational instability mainly induces the associated nonlinearity. Note that the linear focused wave can also be formed with other methods (Onorato et al. 2021; Onorato and Suret 2016; Toffoli et al. 2015).

One snapshot of the spatial profiles of linear and nonlinear focused waves is illustrated in Fig. 2.25. Although the same crest height is achieved in the experiments, the spatial

Table 2.10 Parameters of linear and nonlinear focused waves

Case	LFW1	LFW2	LFW3	LFW4	NFW1	NFW2	NFW3	NFW4
Crest (cm)	6.3	8.8	10	11	6.3	8.8	10	11
Period (s)	0.98	1.18	1.2	1.3	0.98	1.08	1.2	1.3
Steepness	0.26	0.255	0.28	0.27	0.26	0.30	0.28	0.27
Relative depth	4.19	2.91	2.81	2.42	4.19	3.46	2.81	2.42

Fig. 2.25 Spatial profiles of linear (left) and nonlinear (right) focused waves

profiles of these two waves are different at the focusing moment. The crest of the linear focused wave (left) tends to curl over forward, which cannot be identified in the nonlinear focusing process. The spatial profile of nonlinear focused waves is narrower than that of linear focused waves, which is in agreement with the previous numerical simulation result.

However, in the time domain, the profile discrepancy is insignificant, as shown in Fig. 2.26. Except for case 2 (denoted by LFW2 and NFW2), where the trough-to-trough periods are different, the high correlation coefficients of linear and nonlinear focused waves in the other three cases quantitatively verify that they are almost identical in wave profile. Note that even though the crest height is gradually increased in these four cases, the variation of wave nonlinearity is relatively small, mainly in the range from 0.255 to 0.30, which is set with the purpose of avoiding the influence of wave breaking (Alberello et al. 2018; Shemer and Ee 2015).

In addition, it can be observed that the linear focused wave (solid curve in the shaded area) is characterized by a narrower profile compared with the surrounding individual waves within the same wave group (Slunyaev 2018). Thus, it can be confirmed that to a certain degree, the irregular individual wave in Fig. 2.24 is focused, merely measured not at the exact focal location. This can also be proved by checking the phase of each frequency component. Nevertheless, the nonlinearity is still strong enough to dominate the wave evolution at this stage. It thus can suppress the influence of the dispersive effect on the dynamic pressure, consequently leading to much smaller dynamic pressure than those in the common irregular waves.

For all focused waves, the normalized maximum dynamic pressure per unit crest height is plotted in Fig. 2.27. The nonlinear focused waves present a larger maximum dynamic pressure than the linear focused waves. However, if the wave nonlinearity (i.e., longer trough-to-trough wave period in LFW2) is weaker, the maximum dynamic pressure in the linear focused wave can also be greater than that in the nonlinear focused wave with the identical crest height, as revealed by the two points in the green circle in Fig. 2.27. This is consistent with the former conclusion drawn from the regular and irregular individual waves that maximum dynamic pressure is inversely proportional to the wave nonlinearity.

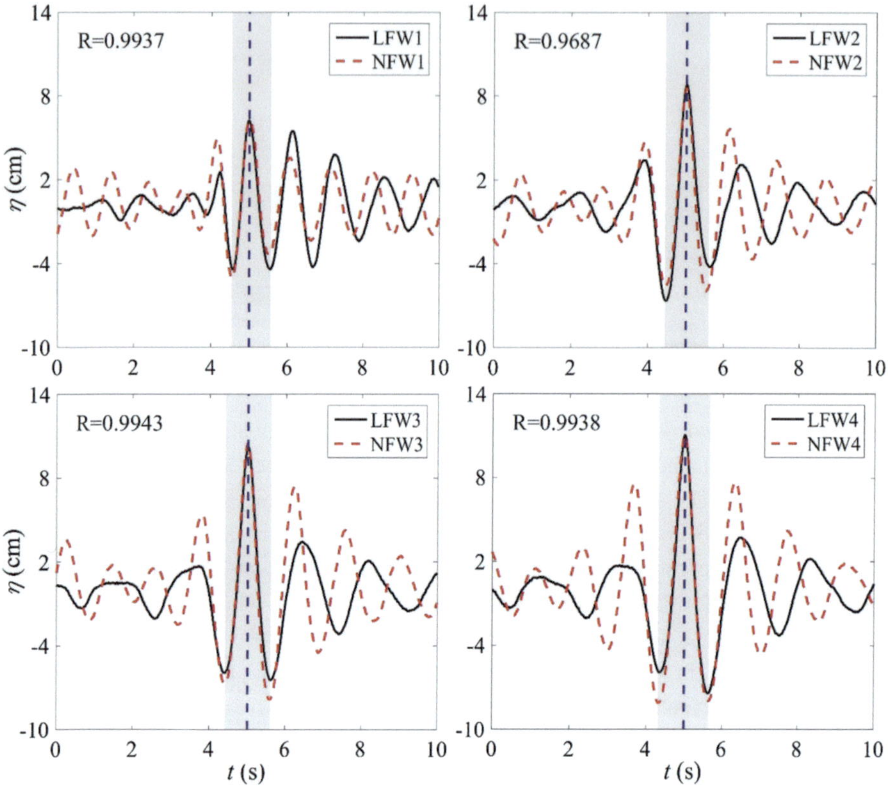

Fig. 2.26 Time series of focused waves generated by different mechanisms

For each type of focused wave, an overall decreased tendency can be detected between the normalized maximum dynamic pressure and the wave nonlinearity, indicated by the steepness listed in Table 2.10.

The vertical profiles of dynamic pressure under focused wave crests are shown in Fig. 2.28, where the maximum value (excluding Case 2) tends to appear at the mean water level in the nonlinear focused waves, higher than that presented in the linear focused waves. Moreover, in the upper section of the vertical profile, the dynamic pressure is always more significant under the nonlinear focused crest, while in the lower section, this is reversed (Klein et al. 2016). Therefore, the linear focused wave tends to induce larger wave loads on marine structures since the location of wetted surfaces of the floating body mostly corresponds to the lower section of the pressure profile. Considering that the wavelength and crest height are identical in this study, it can be inferred that the difference in dynamic pressure is mainly induced by the distinct wave nonlinearity in the focusing process.

Fig. 2.27 Normalized maximum dynamic pressure under the linear and nonlinear focused crests

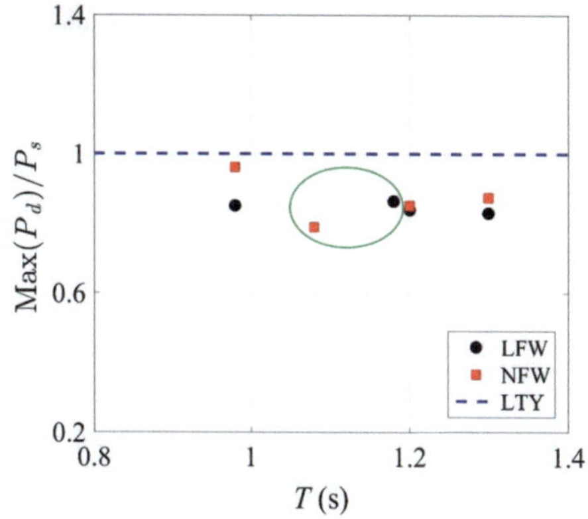

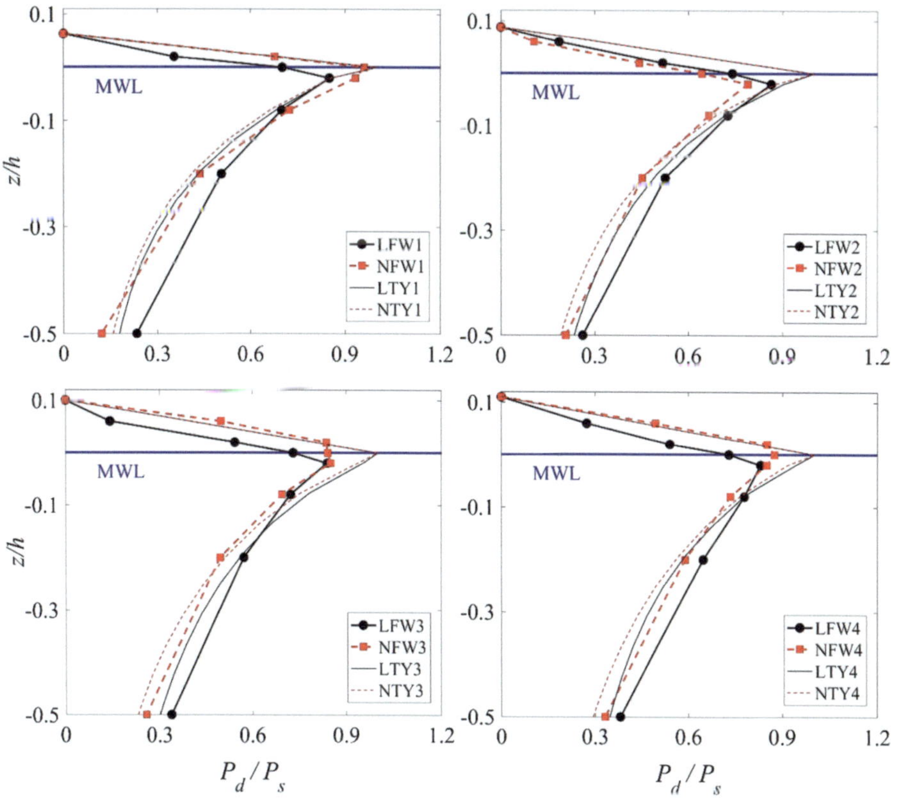

Fig. 2.28 Comparison of dynamic pressure under the linear and nonlinear focused crests. 'LTY' and 'NTY' denote the linear theoretical predictions for linear and nonlinear focused waves, respectively

Moreover, for the linear focused waves, the pattern of deviation from linear theoretical predictions is consistent with that observed in Fig. 2.24 (black circle) but different from those observed in the general irregular waves. Therefore, it can further confirm that the local wave focusing indeed occurred within the wave group depicted in Fig. 2.24 and that the effect of bound-wave nonlinearity on the dynamic pressure almost reached its maximum due to the phase coherence of each wave component, thus resulting in very small pressure in the linear focused wave.

Interestingly enough, for most nonlinear focused waves, the dynamic pressure above the mean water level can be predicted reasonably well by the linear wave theory. However, in Case 2, the dynamic pressure suddenly becomes very small compared to linear theoretical predictions. Note that the steepness of nonlinear focused waves, in this case, is as large as 0.30, meaning that the maximum amplification factor of the breather solution is almost achieved at this location (Waseda et al. 2019). Thus, it can be inferred that at different evolutionary stages, the dynamic pressure in the nonlinear focused waves governed by modulational instability can be largely different, especially noticeable in the region above the mean water level.

To be specific, at the early stage, the nonlinearity of the approximately monochromatic breather solution is dominated by bound wave components, and thus, the dynamic pressure will be smaller than the linear theoretical prediction, following the rule derived from regular waves. The carrier wave is disturbed at the intermediate stage, and the quasi-resonant four-wave interaction dominates the evolutionary process (Slunyaev and Dosaev 2019). Consequently, the third-order nonlinearity will gradually increase the dynamic pressure under the crest, fitting reasonably well with the linear theoretical prediction. At the later stage, once the maximal amplification factor of the Peregrine breather solution is almost reached, especially close to 3, the dynamic pressure will quickly decrease, finally smaller than the linear theoretical prediction once again.

2.6 Evolutionary Property of Multidirectional Waves

To better reveal the spatial variations of characteristics of extreme waves in multidirectional waves and indicate how different nonlinear effects work together on the whole evolutionary process, two different types of experiments identified by the initial Benjamin-Feir index are systematically analysed in this section in the aspects of spectral and statistical properties (Zhang et al. 2019).

The main parameters of the multidirectional waves, characterized by the JONSWAP power spectrum, are listed in Table 2.11. Detailed descriptions of experimental setup can be found in the original works (Onorato et al. 2009) and thus are omitted here for brevity.

Moreover, five different values of the spreading coefficient N (see Fig. 2.29) have been used to model the energy distribution in the directional domain to consider the influence of the directional spreading effect. Note that Janssen and Bidlot (2009) proposed a new

Table 2.11 Parameters of multidirectional waves in laboratory experiments

Case	H_s(m)	T_p (s)	α	ε	γ	*BFI*
A	0.06	1.0	0.014	0.13	3.0	0.70
B	0.08	1.0	0.016	0.16	6.0	1.10

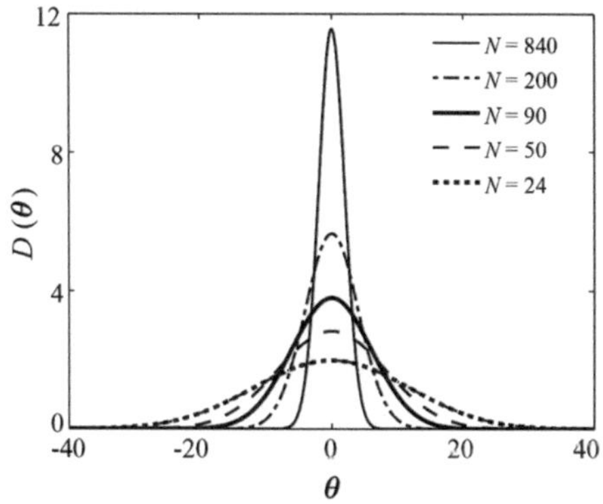

Fig. 2.29 Angular distribution of the initial wave energy. Large *N* corresponds to long-crested waves, while small *N* indicates short-crested waves

definition for *BFI,* which could include information on directional spreading. Still, the original one proposed by Onorato et al. (2001) is used in this book, which is more suitable for decoupling the influence of directionality and nonlinearity on the evolutionary process of extreme waves.

2.6.1 Directional Spectrum

Wave spectrum can show the distribution of wave energy among different wave frequencies or wavelengths. Thus, it should be able to indicate the formation of extreme waves induced by nonlinear wave focusing.

However, for unidirectional waves, after analysing some laboratory experiments conducted in different wave tanks, it is found that the wave spectrum does not always present very prominent variation in this process except for the slightly decreased slope of the equilibrium range (e.g., Fedele et al. 2010).

Nevertheless, it does not mean the energy exchange does not happen in the nonlinear focusing process. It is not easy to be visible in the sea state characterized by a continuous broad wave spectrum (e.g., in the case of $\gamma = 3$). This argument can be substantiated by observing the evolution of various breather solutions, such as Peregrine or Kuznetsov-Ma breather, of which the major wave components are identified easily and clearly. The amplitude of their carrier wave decreases sharply with apparent energy exchange among

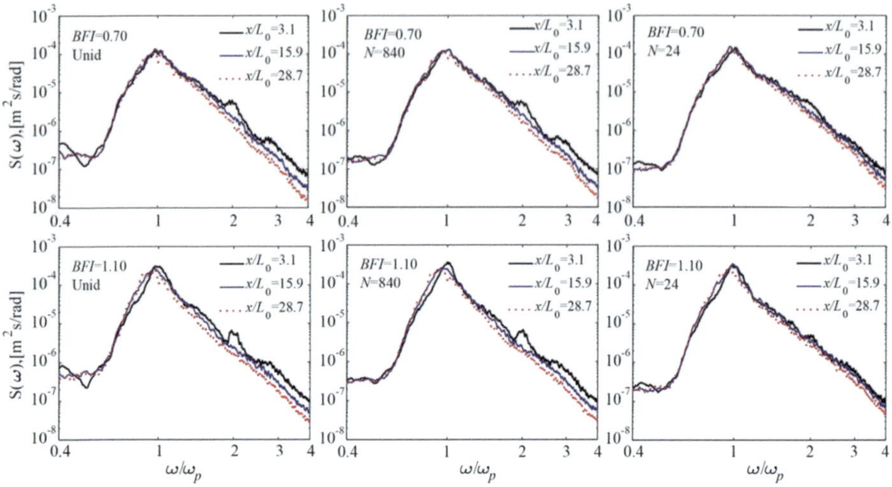

Fig. 2.30 Spatial variations of wave spectrum for unidirectional ("Unid") and directional ($N = 24$ and $N = 840$) water waves

different wave components when the modulational instability is triggered (Shemer and Alperovich 2013; Chabchoub et al. 2017; Zhang et al. 2017a).

In Fig. 2.30, the spectrum in case A ($BFI = 0.70$) and case B ($BFI = 1.10$) exhibits a similar evolutionary tendency in the presence of the same directional spreading coefficient N. The downshift of peak frequency and the decrease of the slope of the equilibrium range are both detected with various degrees, mainly depending on the directional spreading effect.

However, the mechanisms involved may be quite different. As revealed in the previous research (e.g., Trulsen and Dysthe, 1997; Dias and Kharif, 1999), such kind of spectral variations can be attributed to a lot of factors, for example, nonlinear dynamics of free waves, wave breaking and dissipation, as well as oblique sideband perturbation in directional sea state.

Moreover, it has to be mentioned that in two dimensions, without any dissipative effects added, there is no permanent frequency downshift, while the situation will be different in three dimensions. Besides, another interesting phenomenon is that as the directional spreading effect increases, the bound wave components (e.g., $2\omega_p$) at the location of $x/L_0 = 3.1$ gradually decrease.

In Fig. 2.31, three examples of cases A and B are given to illustrate the evolutionary tendency of directional spreading of wave energy. The obvious broadening trend along the basin will decrease the chance of concentrating wave energy into a small area and consequently suppress the formation of rogue waves.

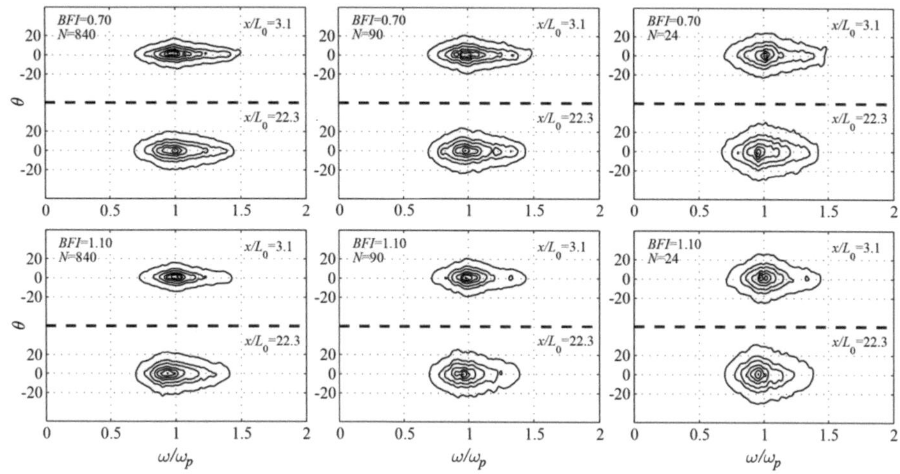

Fig. 2.31 Variation tendency of the directional spectrum

2.6.2 Higher-Order Statistics

The coefficient of skewness (λ_{30}) can indicate the vertical asymmetry of the wave profile. Even in the frame of third-order wave theory, it is still mainly provoked by bound wave effects.

As shown in the first column of Fig. 2.32, the coefficient of skewness presents a slightly decreasing trend along the wave basin, which is observed in all cases. It is mainly due to the decreased amplitude of bound waves, which can be inferred from the reduced level on the right tail of the frequency spectrum. Moreover, the contingent downshift of peak frequency can further reduce the wave steepness. Consequently, the bound wave effects are weakened, reducing the vertical asymmetry of the wave profile.

Based on Gram–Charlier (GC) series approximation, Tayfun and Fedele (2007) proposed the third-order nonlinear GC model for long-crested waves, which will be consistent with the MER model of Mori and Janssen (2006) if $\Lambda \to \Lambda_{app} = 8\lambda_{40}/3$ where $\Lambda = \lambda_{40} + 2\lambda_{22} + \lambda_{04}$ and λ_{ij} represents the fourth-order normalized joint cumulants. According to past research on 2D extreme waves from different laboratory experiments and numerical simulations performed with various wave models, $\Lambda \to \Lambda_{app}$ only holds in the absence of vertical asymmetry.

More often than not, Λ will be smaller than Λ_{app}. This is contrary to the conclusion obtained from the second-order wave theory where $\lambda_{40} < 3\lambda_{22} < \lambda_{04}$ (Tayfun and Fedele 2007). The first column of Fig. 2.32 shows that $\lambda_{40} > \lambda_{04} > 3\lambda_{22}$, and their difference will be enlarged as the vertical asymmetry of the wave profile increases. To a certain degree, it is consistent with the theory developed by Janssen and Bidlot (2009), who pointed out that the total λ_{40} consists of the sum of the "dynamics" contribution and the "wave

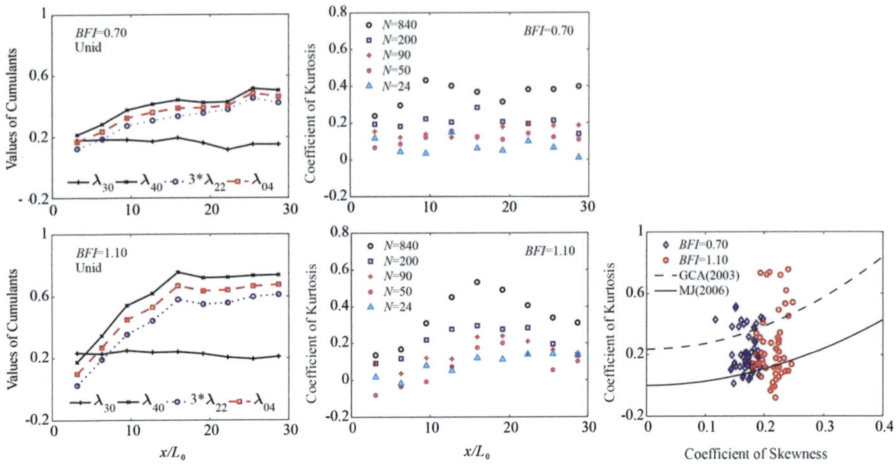

Fig. 2.32 Relationship between third-order and fourth-order cumulants

shape" contribution. Therefore, for long-crested nonlinear waves, the relationship among fourth-order joint cumulants has to be discussed in the context of both second-order and third-order nonlinear wave theories.

However, if the directional spreading effect is considered, these fourth-order cumulants, for example, the coefficient of kurtosis presented in the second column of Fig. 2.32, will be dramatically changed, while the third-order cumulants such as the coefficient of skewness only show little variation with the directional spreading coefficient *N*.

In the sea state with large initial *BFI* and *N*, the whole wave field in the basin is inhomogeneous, and its fourth-order statistics vary largely along the basin. Consequently, it will be very difficult to establish the relationship between coefficients of kurtosis and skewness (see the third column of Fig. 2.32). In other words, those quadratic functions (Guedes Soares et al. 2003; Mori and Janssen 2006) can fit the experimental results reasonably well only if the directional spreading effect mostly quenches the nonlinear wave focusing.

2.6.3 Surface Elevation Probabilities

For multidirectional waves, especially in the case with a broad directional distribution (e.g., $N = 24$), it makes no sense to discuss the evolutionary stages, considering that the dynamics of free waves no longer significantly contribute to the statistical properties of extreme waves. Strictly speaking, the crest and trough of multidirectional waves are also very difficult to accurately determine from the recorded time series at one point, not to mention the height and period (Dysthe et al., 2008).

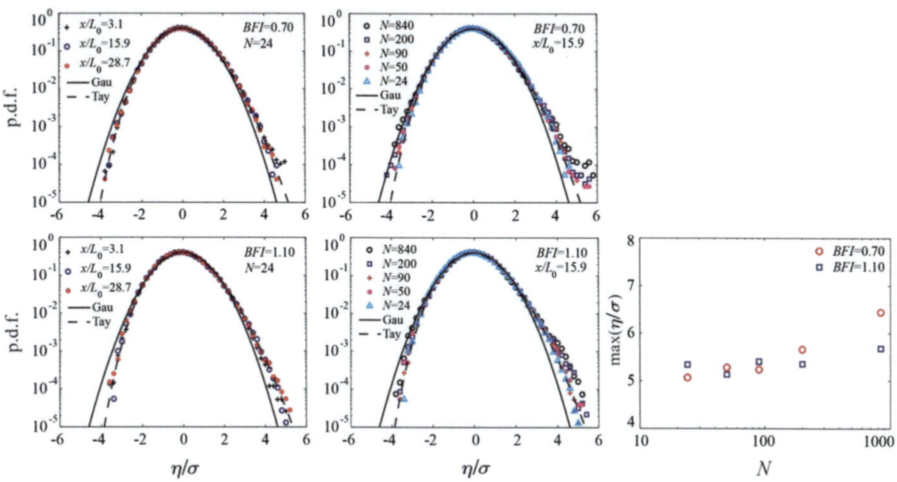

Fig. 2.33 Surface elevations for multidirectional waves

Therefore, in Fig. 2.33, only surface elevations are analyzed for multidirectional waves. To a certain degree, they can implicitly indicate the probability of observing extreme waves in the basin. Apparently, for large directional spreading (e.g., $N < 100$), the probability density function of surface elevations in the whole basin is fitted very well by Tayfun distribution regardless of the initial *BFI*. Therefore, the second-order theory can predict the probability of observing extreme waves in these cases.

However, as waves become long crested, the maximum of the scaled surface elevation grows quickly with N, particularly pronounced in the case with a smaller initial Benjamin-Feir index. Note that in the third column of Fig. 2.33, all dimensional quantities in case B (blue squares) are larger than those in case A (red circles). The appearance of extreme waves can be underpredicted by about one order of magnitude in the frame of second-order wave theory.

2.7 Conclusion

The statistical wave parameters present similar spatial variations for the same sea state in two wave basins with different aspect ratios. The wave frequency spectrum measured in CEHIPAR is always a little narrower than that obtained in Marintek, which can be attributed to the different geometry of the wave basins. The difference in the exceedance distribution of wave height is not conspicuous up to the probability level of 0.1. The wave period density shows the same distribution law determined by the spectral width and other parameters, such as wave steepness.

Two modified versions of Longuet-Higgins joint distribution of wave heights and periods have been proposed. The first one can catch the location of the mode of distribution in all sea states, although some of the contour lines are enlarged in comparison with the observed statistics. The second one is an alternative form of Longuet-Higgins joint distribution. It cannot fundamentally eliminate the disparities with experiments. Still, it is theoretically possible to use it to model the joint distribution of each component in a mixed sea state and shed some light on how to further extend this kind of theoretical model to a nonlinear wave time series.

Regarding regular waves, the bound-wave nonlinearity has an opposite effect on the upper (decrease) and lower (increase) sections of the dynamic pressure under the wave crest, resulting in different types of deviation from the linear theoretical prediction, particularly pronounced in the case of deepwater waves. Besides, the location of maximum dynamic pressure appears closely below the mean water level in all cases, slightly lower than the theoretical value. Moreover, the growth rate of dynamic pressure under wave crests negatively correlates with wave nonlinearity by changing wave period or wave amplitude.

Concerning irregular waves, from a statistical point of view, the dynamic pressure under wave crests is still inversely proportional to the nonlinearity of individual waves, indicated by the steepness calculated with the trough-to-trough period. However, in most cases, the dispersive effect tends to dominate the evolutionary process, which can strengthen the dynamic pressure to various degrees, leading to vastly different pressure fields for similar individual waves. To consider the influence of dispersive effects simultaneously, the statistical parameter such as peak period is more reliable to work than other indexes to evaluate the corresponding maximum dynamic pressure under similar individual waves, which still appears below the mean water level and is inclined to be larger than its counterpart in regular sea state.

As to the same focused waves generated with different physical mechanisms, the dynamic pressure field under wave crests is entirely different due to the distinct nonlinearity dominant at the focusing moment. For linear focused waves, the dynamic pressure above the mean water level is dropped mainly because the effect of bound-wave nonlinearity reaches its maximum value as a result of the phase coherence of each wave component, overwhelming the contribution of the dispersive effect and thus inducing smaller pressure than that of the common irregular waves. For the nonlinear focused waves, the effect of third-order nonlinearity associated with modulational instability on the dynamic pressure is complex and dependent on the evolutionary stage. Once it dominates the wave evolutionary process, the effect of bound-wave nonlinearity on the dynamic pressure above the mean water level can be completely counteracted at some moments, which is revealed by the excellent fit with the linear theoretical prediction in the upper section of the pressure profile.

For multidirectional waves, the peak frequency's downshift and the decrease in the equilibrium range slope are detected at various degrees, mainly depending on the directional spreading effect. The evident broadening trend along the basin will decrease the chance of concentrating wave energy into a small area and consequently suppress the formation of rogue waves. The fourth-order cumulants, such as the coefficient of kurtosis, are dramatically changed, while the third-order cumulants, such as the coefficient of skewness, only show slight variation with the directional spreading coefficient. In the case of a large directional spreading coefficient, the probability density function of surface elevations is fitted very well by the Tayfun distribution regardless of the initial BFI. The second-order wave theory can reasonably predict the probability of observing extreme waves in these cases.

Modelling Extreme Waves with 2D NLS Equation

3

3.1 Basic Theory

3.1.1 Cubic Nonlinear Schrödinger Equation

The simplest weakly nonlinear model that describes the evolution of free waves is the so-called cubic Schrödinger equation, which has been derived from the Zakharov equation under the narrowband approximation and represents a perfect framework in which the basic features of modulational instability are contained. Working as a balance between dispersion and nonlinearity, the dimensional NLS equation in arbitrary depth, in a frame of reference moving with the group velocity, has the following form:

$$\frac{\partial A}{\partial t} + i\alpha \frac{\omega_0}{8k_0^2} \frac{\partial^2 A}{\partial x^2} + i\beta \frac{\omega_0 k_0^2}{2} |A|^2 A = 0, \tag{3.1}$$

where A is the complex wave envelope; ω_0 and k_0 are the carrier wave frequency and corresponding wave number; α and β are two coefficients that generally depend on the dimensionless water depth $k_0 h$, and both tend to 1 as $k_0 h$ approaches infinity. The analytical forms of α and β can be found in many documents, such as the book of Mei (1989), and will not be repeated here. The second term of Eq. (3.1) considers the dispersive behaviour of the surface elevation, while the last one reflects the third-order nonlinear effect.

To indicate the significant effect in the evolutionary process, a parameter named the Benjamin–Feir Index (*BFI*), similar to the "Ursell" number, is introduced (Onorato et al., 2001; Janssen, 2003). To derive the Benjamin–Feir index in a simple and instructive way, Eq. (3.1) is nondimensionalized in the following ways: $A' = A/a_0$, $x' = x\Delta K$ and $t' = t(\Delta K/k_0)^2 \alpha\omega_0/8$, where ΔK represents a typical spectral bandwidth and a_0 is a

H. Zhang and C. Guedes Soares, *Numerical Modelling of Extreme Waves*, Synthesis Lectures on Ocean Systems Engineering, https://doi.org/10.1007/978-3-031-77084-5_3

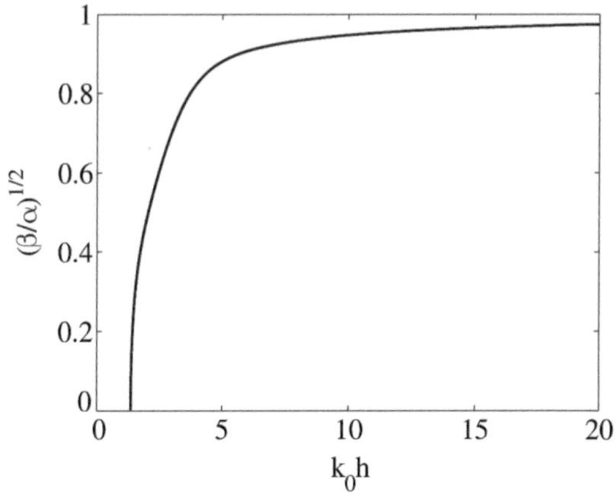

Fig. 3.1 Effect of finite water depth on Benjamin–Feir Index

typical wave amplitude. Thus, it reduces to

$$\frac{\partial A'}{\partial t} + i\frac{\partial^2 A'}{\partial x^2} + i\left(\frac{2\varepsilon}{\Delta K / k_0}\right)^2 \frac{\beta}{\alpha}|A'|^2 A' = 0, \tag{3.2}$$

where $\varepsilon = a_0 k_0$ is a measure of the wave steepness. The Benjamin–Feir Index is now defined as the square root of the coefficient that multiplies the nonlinear term:

$$BFI = \frac{2\varepsilon}{\Delta K / k_0}\sqrt{\frac{\beta}{\alpha}}. \tag{3.3}$$

The term $\sqrt{\beta / \alpha}$ is close to one as $k_0 h$ tends to infinity and decreases as the water becomes increasingly shallow, appearing negative for values of $k_0 h$ smaller than 1.36 as shown in Fig. 3.1. In the latter case, the modulational instability disappears, and Stokes waves are stable to perturbations (e.g., Onorato et al., 2001).

Considering that time series are usually measured in wave tanks or open sea, the term $\Delta K / k_0$ in the BFI is replaced by $2\Delta\omega / \omega_0$ for infinite water depth. Due to historical reasons, a factor $\sqrt{2}$ is also added to Eq. (3.3). Usually, if$BFI \ll 1$, water waves are essentially linear and their dynamics can be approximately expressed as a simple superposition of sinusoidal waves. If $BFI > 1$, , the dynamics become nonlinear, and the evolution of the wave train is likely dominated by envelope solitons or unstable mode solutions (Onorato et al., 2001; Janssen, 2003).

If the initial condition of the surface wave is characterized by a large BFI, the probability of finding extreme waves increases. A more formal way of relating abnormal waves to BFI has been introduced by Janssen (2003). Starting from the Zakharov equation, he concluded that for the long-crested narrowband waves the asymptotic state of the coefficient of kurtosis λ_{40} (for a large time) is strictly related with BFI on a square relationship.

$$\lambda_{40} = \frac{\pi}{\sqrt{3}} BFI^2. \tag{3.4}$$

Considering that *BFI* has a larger variability in its computation, it makes sense to use *BFI* as an initial parameter to indicate the chance of observing the freak waves and to take λ_{40} λ_{40} as a critical parameter to describe the extreme wave height in the observed series.

Based on the NLS equation, Fedele et al. (2010) derived an analytical expression describing the spatial evolution of the coefficient of kurtosis for the stationary waves with a Gaussian-shaped initial spectrum. As Onorato and Suret (2016) pointed out, this is a significant result because it represents the starting point for developing a tool to forecast extreme waves operationally.

Considering that nowadays, the forecasting of ocean waves is based on the numerical integration of the energy balance equation in spectral space, it makes sense to forecast the extremes according to some spectrum properties. Once an estimation of the wave spectrum is made, it is possible to compute the *BFI* and, consequently, the coefficient of kurtosis.

3.1.2 Theoretical Distribution Models of Wave Heights

The statistics of unusually large waves generated by modulational instability can be explained reasonably well by the theoretical approximations based on Gram–Charlier (GC) expansions (Tayfun and Fedele 2007; Mori and Janssen 2006). Such approximation represents Hermite series expansions of distributions describing non-Gaussian random functions, which are related to the stochastic structure of waves only through certain key statistical variables such as the coefficient of kurtosis of surface elevation.

Mori and Janssen (2006) proposed a modified Edgeworth–Rayleigh (MER) distribution for narrowband long-crested waves. It should be stressed that MER assumes weakly nonlinear random waves with narrowband spectrum and further requires specific relationships between the normalized fourth-order joint cumulants λ_{40}, λ_{04} and λ_{22}, i.e., $\lambda_{40} = \lambda_{04}$ and $\lambda_{40} = 3\lambda_{22}$ in comparison with the more general representation form (GC Model), which is given by Tayfun and Fedele (2007):

$$E(h) = \exp\left(-\frac{h^2}{8}\right)\left[1 + \frac{\Lambda}{1024} h^2\left(h^2 - 16\right)\right], \tag{3.5}$$

where $\Lambda = \lambda_{40} + 2\lambda_{22} + \lambda_{04}$ accounts for the third-order quasi-resonant nonlinear interaction effect.

Only in the special condition: $\Lambda \rightarrow \Lambda_{app} \equiv 8\lambda_{40}/3$, GC model converges to MER model

$$E(h) = \exp\left(-\frac{h^2}{8}\right)\left[1 + \frac{\lambda_{40}}{384}h^2(h^2 - 16)\right]. \tag{3.6}$$

The cumulant coefficients in Eq. (3.5) are expressed in the following form:

$$\lambda_{mn} = \langle \eta^m \hat{\eta}^n \rangle / \sigma^{m+n} + (-1)^{m/2}(m-1)(n-1), \tag{3.7}$$

where η is the surface elevation; $\hat{\eta}$ is its Hilbert transformation; the bracket $\langle \cdot \rangle$ represents statistical averaging, and σ designates the standard deviation of the surface displacement.

For linear waves, $\Lambda \cong 0$ and consequently Eq. (3.5) reduces to the normalized Rayleigh exceedance distribution, that is,

$$E(h) = \exp\left(-\frac{h^2}{8}\right). \tag{3.8}$$

3.1.3 Numerical Schemes

The NLS equation's numerical simulation has been performed using a standard split-step pseudo-spectral Fourier method (Lo and Mei 1985). The nonlinear part is integrated by a finite difference method in physical space, while the linear dispersive part is solved exactly in the Fourier space.

For the more realistic ocean environment, the initial condition in the numerical simulation is a typical sea state described by the JONSWAP power spectrum:

$$S(\omega) = \alpha_1 g^2 \omega^{-5} \exp\left[-\frac{5}{4}\left(\frac{\omega_0}{\omega}\right)^4\right] \gamma^{\exp\left[-(\omega-\omega_0)^2/\left(2\sigma_0^2\omega_0^2\right)\right]}, \tag{3.9}$$

where $\sigma_0 = 0.07$ if $\omega \leq \omega_0$ and $\sigma_0 = 0.09$ if $\omega > \omega_0$. Here ω_0 and γ represent the peak frequency and peak enhancement parameter with α_1 the Phillips' constant related to the significant wave height H_s. As γ increases, the spectrum becomes higher and narrower around the peak frequency. In the present simulations $\gamma = 3$, $\varepsilon = k_0 H_s/2$ and $\Delta\omega$ is estimated with half-width at half maximum of the computed wave spectrum (Onorato et al., 2006). One must remember that *BFI* and ε will reflect the same initial condition if the peak enhancement parameter γ is the same among all sea states.

The initial condition for the free surface elevation $\eta(0, t)$ can be synthesized by the following random process:

$$\eta(0, t) = \sum_{n=1}^{N} C_n \cos(\omega_n t - \phi_n), \tag{3.10}$$

where ϕ_n are uniformly distributed random numbers in the interval $[0, 2\pi)$, and $C_n = \sqrt{-2S(\omega_n)\ln(u_n)\Delta\omega_n}$ with u_n being another independent random number uniformly distributed in the range $[0, 1)$ to produce random amplitudeC_n. Immediately, the complex envelope $A(0, t)$ can be derived by a further Hilbert transform of the free surface elevation.

The numerical simulation solves a boundary value problem: given the temporal variation $A(0, t)$ at some location $x = 0$, the governing equation determines the wave motion over all space and time, $A(x, t)$.

In the computational domain concerned, it has been noted that the shape of the JONSWAP spectrum has not been substantially modified during the evolution of the NLS equation. The time series of the free surface at a specified location can be obtained to the leading order by the following expression:

$$\eta(x, t) = \{A \exp[i(k_0 x - \omega_0 t)] + c.c.\}/2, \tag{3.11}$$

where $c.c.$ denotes the corresponding complex conjugate function.

3.2 Typical Wave Envelopes

In this section, two simple but typical wave envelopes are analyzed in the following numerical tests to better understand the capability of the NLS equation in describing the evolution of a propagating wave packet with a narrow spectrum.

The first simulated series is a Gaussian wave packet, and its initial surface elevation is:

$$\eta(t) = a_0 \exp\left[-\left(\frac{t}{mT_0}\right)^2\right] \cos(\omega_0 t), \qquad -16T_0 < t < 16T_0, \tag{3.12}$$

where the carrier wave period $T_0 = 2\pi/\omega_0$. The energy spectrum of Eq. (3.12) also presents a Gaussian shape and its relative width at half maximum is given by $\Delta\omega/\omega_0 = \sqrt{2\ln 2}/(2m\pi)$. The value of the parameter m is set to be 4.0 in this chapter so that all cases considered meet the condition $\Delta\omega/\omega_0 = 0.047 < \varepsilon$, thus satisfying the narrowband spectrum assumption of the NLS equation.

In the second kind of numerical test, the bichromatic wave has been studied with the initial surface elevation in the following form:

$$\eta(t) = a_0 \cos(\omega_0/20t)\cos(\omega_0 t), \qquad -15T_0 < t < 15T_0. \tag{3.13}$$

The carrier wave frequency and maximum amplitude of the bichromatic wave will be identical to those used in Eq. (3.12) in the following study. This kind of surface elevation spectrum is bi-modal, with two equal-height peaks at $\omega = \omega_0 \pm \Delta\omega$, where $\Delta\omega = \omega_0/20$, satisfying the requirement of narrowband spectrum approximation again (Shemer et al. 2002).

The parameters used in the envelope analysis are assigned in purpose from Table 2.4, the forcing amplitude $a_0 = H_s/2$ and the carrier frequency $\omega_0 = 2\pi/T_p$.

The initial wave envelopes are derived by Hilbert transform (Veltcheva et al. 2016) of the surface elevations expressed in Eqs. (3.12) and (3.13) and their amplitudes are depicted on the two columns in Fig. 3.2, respectively.

The evolutionary envelopes are normalized by their corresponding maximal wave amplitudes in the initial conditions. For the economy of space, only three results from Cases 4, 13 and 23 in Table 2.4 are presented from top to bottom in Fig. 3.2, where the spatial evolutionary tendencies of envelopes are typical and representative since they belong to three different sea states.

In the low sea state, e.g., Figs. 3.2a, d, no significant variation of the wave envelope along the tank can be observed either for the Gaussian wave packet or for the bichromatic wave group for the reason that the modulational instability is weak as indicated by the small initial steepness.

In the moderate sea state, e.g., Fig. 3.2b, e, they show similar evolutionary speed and tendency. Take the single envelope pulse as an example, the initial Gaussian wave group adjusts its shape and width to become a fundamental soliton with oscillatory tails and then attains a "sech" profile. Moreover, concerning the evolution of a nonlinear continuous wave train, the amplitude does not grow exponentially for all time. Instead, it grows to a maximum value and then decreases in magnitude for later times, repeating this oscillation periodically over time (Fermi–Pasta–Ulam or FPU recurrence). This modulation and demodulation process has also been verified in the experiments, and this kinds of oscillatory unstable modes are often referred to as breathers.

In the severe sea state, e.g., Fig. 3.2c,f, the same variation can be observed but with a higher evolutionary speed and a steeper envelope due to the stronger modulational instability. In the presence of stronger nonlinearity for a very long evolutionary distance, both Gaussian and bichromatic wave envelopes can display a periodic variation in numerical simulations, more or less like the behaviour of continued series, i.e., FPU recurrence.

Last, the symmetry of the initial condition is conserved in all cases, which contradicts reality, but the envelope modification is quantitatively correct and has been confirmed many years before. Thus, the NLS equation can be applied to the simulation of extreme waves since it can describe the evolution of nonlinear wave packets.

3.3 Simulation of Laboratory Experiments in Marintek

The discussions in this section are based on the results obtained from four different experiments listed in Table 2.2, allowing the comparison of three typical sea states with the same series length: smooth (8201, triangle), moderate (8202, square) and severe (8219,

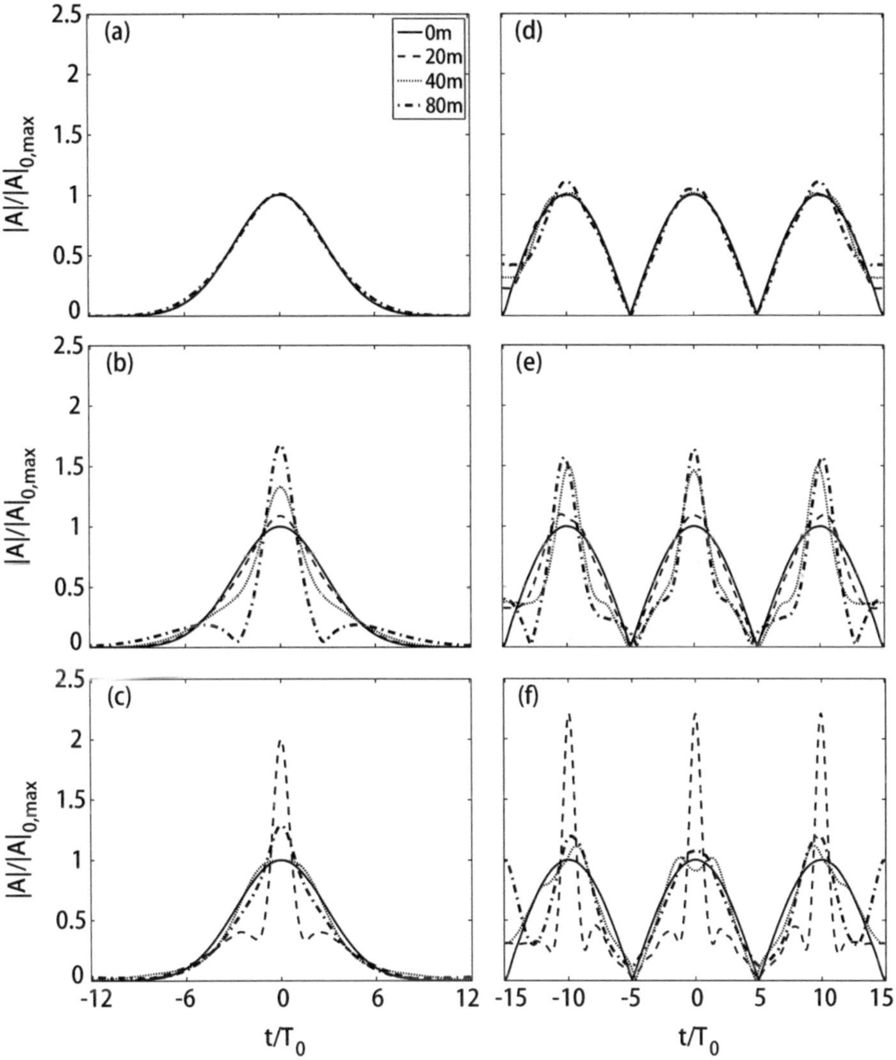

Fig. 3.2 Spatial variations of wave envelopes in three typical sea states. **a–c** are from Gaussian wave packets and **d–f** are from bichromatic wave groups. The three rows correspond to the smooth, moderate and severe sea states

diamond), and also the comparison in the moderate sea states with different series lengths (8241–45, circle). For clarity, different symbols are used to identify different sea states, and the full and empty marks denote experimental and simulated results, respectively, in the following figures.

3.3.1 General Comparisons

The relationship between the coefficient of kurtosis and *BFI* is plotted in Fig. 3.3. As expected, the theoretical function (solid line) represented in Eq. (3.4) overestimates the results in experiments and simulations for the reason that the analytical expression is derived for the nonlinear steady state at the infinity while the data points are from all wave gauges, for most of which the nonlinearity is not fully developed. Moreover, in laboratory experiments, the coefficient of kurtosis is more sensitive to wave breaking in high sea states than *BFI* (see their definitions), which definitely leads to a further deviation from the theoretical prediction as we can see (the full diamond symbols) in Fig. 3.3. However, from a point of statistical view the general tendency is correct, that is, the series with larger *BFI* is also accompanied with a larger coefficient of kurtosis. Furthermore, after a detailed observation, it can be concluded that the coefficient of kurtosis almost maintains a stable, relatively constant value if $BFI > 1.0$.

Tomita and Kawamura obtained an appreciable correlation between the scaled maximum wave height and the kurtosis coefficient based on many abnormal waves registered in the Japan Sea. Guedes Soares et al. (2003) gave a linear regression model after analyzing all sea state records during the storm of November 1997 in North Alwin and Draupner's storm at the beginning of 1995. As shown in Fig. 3.4, a similar conclusion can be clearly seen in the present experiments and simulations.

Meanwhile, since the maximum wave height (empty circle in simulation and full circle in experiment) appears in the case of 8241–45, it strongly confirms the existing conclusion that the extreme wave height is not only determined by the coefficient of kurtosis but also related to the number of waves in the concerned series (Mori and Janssen, 2006). Note that the results in the simulations are a bit larger than those in the experiments, particularly in the case with a larger initial *BFI*. This could be attributed to the non-breaking phenomenon, which leads to no energy dissipation and the lack of higher-order

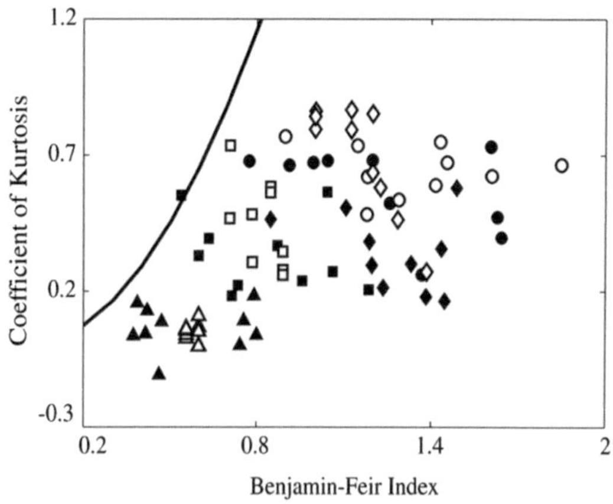

Fig. 3.3 Relationship between the coefficient of kurtosis and *BFI* in Marintek dataset

Fig. 3.4 Relationship between scaled maximum wave height and coefficient of kurtosis in Marintek dataset

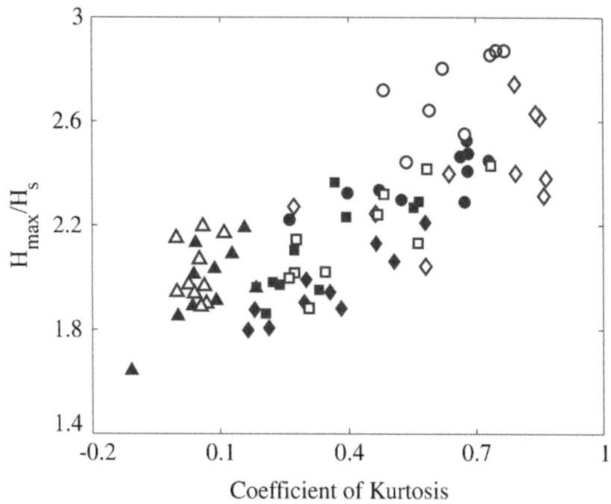

nonlinear dispersive terms, thus permitting the formation of larger amplitude waves in numerical simulation.

In Fig. 3.5, it can be seen that Λ is almost equal to Λ_{app} in all sea states in the simulation while this is not true in the experiment except for the case with small initial steepness, which is depicted by the full triangles. In general, Λ is smaller than Λ_{app} and the difference will be slightly enlarged as the wave steepness increases, maybe due to the Stokes contribution (Cherneva et al., 2013). Furthermore, large Λ appears in the sea state with large initial *BFI*. Since the values of Λ derived in 8241–45 (full and empty circles) are larger than those in 8202 (full and empty squares), it can be concluded that more realizations are necessary to obtain a stable estimation and that much longer relative evolutionary distance is needed to allow the third-order nonlinearity to be fully developed.

To clearly indicate the behaviour of some statistical quantities, such as λ_{40} can work as an indicator of the presence of extreme events in the time series (Guedes Soares et al.,

Fig. 3.5 Relationship between Λ and Λ_{app} in the Marintek dataset

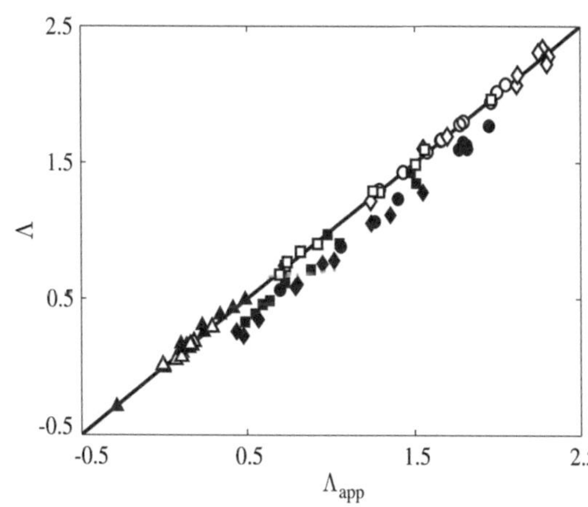

2004), the evolution of the coefficient of kurtosis in Fig. 3.6 and the spatial variation of rogue wave density in Fig. 3.7 will be compared together. Due to the limited number of ensembles in the experiment, only wave series of the 8241–45 redord can be analyzed. The density of rogue waves is defined as the ratio of the number of wave heights, including both zero up-crossing and down-crossing waves, which satisfy the condition $h > 2H_s$, to the total number of recorded wave heights. The distance has been scaled by the dominant wavelength, $L = 2\pi g/\omega_p^2$ corresponding to the initial peak frequency ω_p.

The evolutionary tendency of the coefficient of kurtosis is in agreement with that of rogue wave density in most cases, not only in experiments but also in simulations. In this sense, the coefficient of kurtosis is thought to be the representative statistical measure for the increase of the total large peak-to-trough wave height and, consequently, is confirmed with respect to the probability of occurrence of abnormal waves.

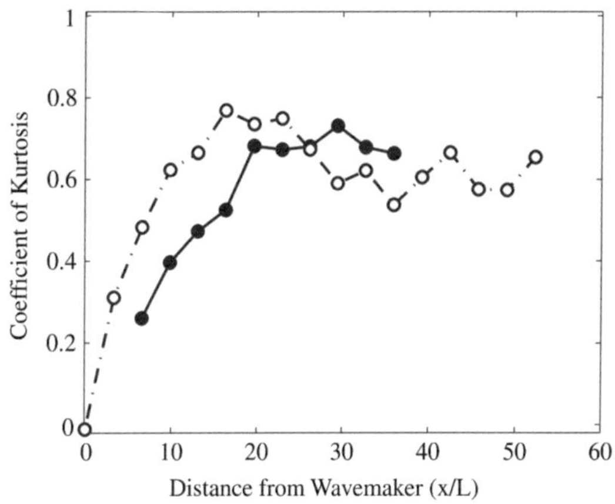

Fig. 3.6 Spatial variation of coefficient of kurtosis in the 8241–45 record

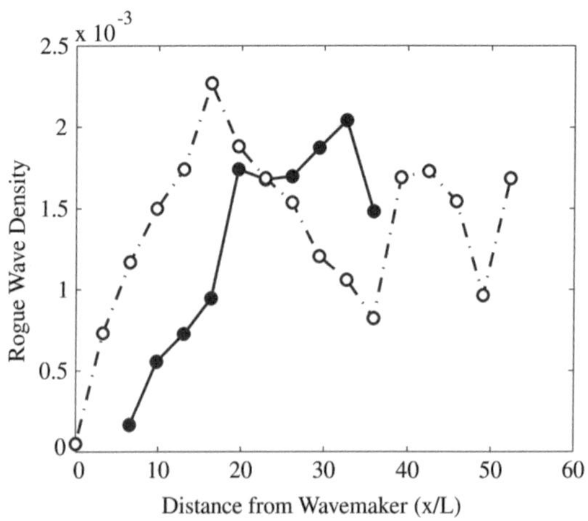

Fig. 3.7 Spatial variation of rogue wave density in the 8241–45 record

Furthermore, compared with the experiment, it can be said that the NLS equation almost gives a reliable simulation of reality, although it reaches the fully developed condition earlier due to no energy dissipation and no higher-order dispersive effect in the numerical process. A similar high evolutionary speed of λ_{40} for unidirectional waves was also detected by Toffoli et al. (2010a) in numerical simulations performed with another two numerical simulation models.

The spatial variation of the coefficient of kurtosis, contributed only by free wave modes, was presented by Fedele et al. (2010). Comparing with the result in Fig. 3.6, it indicates that the large kurtosis is mainly attributed to the quasi-resonant four-wave interaction that takes place near the peak of the spectrum rather than due to the phase-locked modes (Stokes contribution) of the dominant wave numbers with higher harmonics (Onorato et al., 2005; Fedele et al., 2010).

Besides, the range of abnormal wave density is bounded between 10^{-3} and 10^{-4}, which is in conformity with the results of experiments made by Onorato et al. (2006) and those of full-scaled field data analyzed by Guedes Soares et al. (2003). In the context of linear wave theory, the probability of the formation of an extreme wave, derived by Eq. (3.8), is 3.35×10^{-4}, which underestimates most of the experimental and numerical results.

3.3.2 Comparison of Wave Height Distribution

To have sufficiently stable statistics, the exceedance probability of wave heights presented in Fig. 3.8 is based on zero up-crossing and down-crossing waves. For economy of space, the comparison will only focus on the measurements at gauges 1, 5 and 8 corresponding to the three columns in Fig. 3.8.

The sea states become increasingly severe from top to bottom, having the same sequence as those listed in Table 2.2. Moreover, the exceedance distributions are compared with the theoretical predictions of the linear Rayleigh distribution in Eq. (3.8) and the third-order GC model in Eq. (3.5).

From Fig. 3.8a–c, the numerical simulation agrees almost perfectly with the experiment. Since the initial steepness is small, the nonlinearity is negligible, and the wave surface elevation is approximated to be Gaussian distributed. As a result, the GC model tends to linear Rayleigh statistics along the wave tank as expected.

As shown in Fig. 3.8d–f, the NLS equation also reasonably simulates the experiment in the moderate sea state. In Gauge 1 and 5, the wave height is Rayleigh distributed because the nonlinear instability is not very strong at the initial and intermediate stages. As the wave propagates to Gauge 8, nonlinearity has significantly influenced the evolutionary process, and hence, the third-order GC model perfectly fits the larger wave height distribution.

In another moderate sea state, from Fig. 3.8g–i , it is possible to consider the influence of the number of waves on the estimation of extreme wave height because there

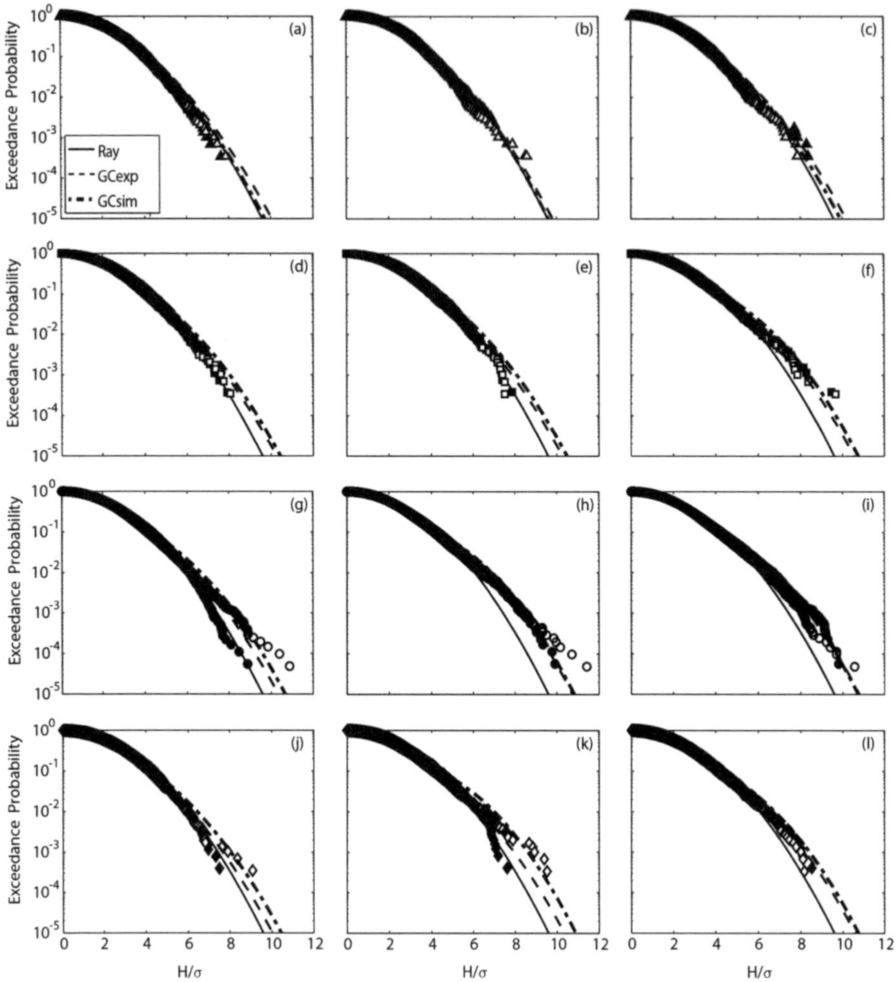

Fig. 3.8 Exceedance distributions of scaled wave heights in the Marintek dataset. The four rows correspond to four different cases listed in Table 2.2. The three columns present the results obtained in the Gauges 1, 5 and 8 locations, respectively

are five similar series at each location. The simulation presents a significant deviation from the experiment at Gauge 1, and this discrepancy is due to the quickly developed strong nonlinear effect in the numerical simulation and the wave breaking in laboratory experiments.

Moreover, a large set of wave series can further enlarge the discrepancy, especially on the lowest probability part. The preceding discussion about Figs. 3.6 and 3.7 has already cast some light on this problem. The wave height in the experiment is Rayleigh distributed, while in the simulation, it can only be described adequately by the third-order theoretical model. In Gauge 5 and Gauge 8, the simulation is consistent with the experiment, and the larger wave height can be adapted to the third-order nonlinear model. However, there is still a small deviation in the tail of the distribution, which could be attributed mainly to statistical variability.

As for the severe sea state, from Fig. 3.8j–l, the same conclusion can be drawn: the NLS equation still captures the main statistical properties of extreme wave heights, particularly in Gauge 1 and Gauge 8. The pronounced deviation in the tail of the exceedance distribution in Gauge 5 is distinguished and apparently generated as a result of serious wave breaking in this location, considering the sea state is already very severe.

Finally, the GC model sometimes fails to predict the extreme wave height distribution, especially in the transition stage where the waves evolve from linear to fully nonlinear, e.g., Fig. 3.8e. So far, there is no chance to forecast the extreme waves in a deterministic method because any deterministic model would fail after about 1000 periods of propagation.

3.4 Simulation of Laboratory Experiments in CEHIPAR

To perform a more complete analysis, the laboratory experiments listed in Table 2.4 will be simulated by the NLS equation in this section. Different symbols will be used to identify the three groups of sea states: square (smooth), circle (moderate), and triangle (severe). As before, the full and empty marks denote experimental and simulated results. Furthermore, the bound wave effects will be removed from the wave series to focus on the dynamics of free waves. The detailed procedure can be found in the work of Fedele et al. (2010).

3.4.1 Exceedance Distribution of Wave Heights

As discussed, for the Marintek dataset in Fig. 3.5, Λ should be smaller than Λ_{app} in the nonlinear wave series generated in the laboratory experiments. However, in Fig. 3.9 Λ is almost equal to Λ_{app} in all sea states in simulations and experiments where the bound waves have been removed.

Thus, it can be concluded that the discrepancy between Λ and Λ_{app} is mainly due to the higher-order bound wave effects, which can be enlarged by the effect of modulational instability, i.e., a quasi-resonant four-wave interaction process that takes place near the peak of the spectrum (Onorato et al. 2005).

Since $\Lambda \approx \Lambda_{app}$, the MER model will make no significant difference from the GC model in predicting extreme wave heights. Hence, the MER model is adopted for the study in this section. To have sufficiently good statistics, the exceedance probability of wave

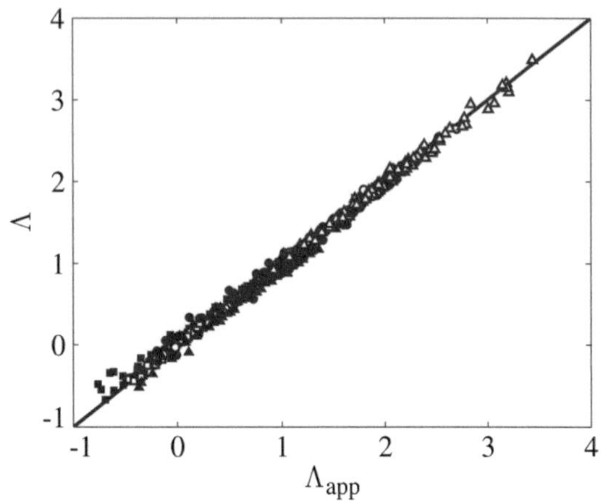

Fig. 3.9 Relationship between Λ and Λ_{app} in the CEHIPAR dataset

heights presented in Fig. 3.10 is also based on the zero up-crossing and down-crossing waves.

Due to the space limitation, the comparison will still focus on three locations, which could represent the initial, intermediate, and peak stages of wave evolution and correspond to the three columns in Fig. 3.10. Three typical cases listed in Table 2.4 are selected to represent the smooth, moderate and severe sea states. Moreover, the exceedance distributions are compared with the theoretical predictions of the linear Rayleigh distribution in Eq. (3.8) and the third-order MER model in Eq. (3.6).

In the smooth or low sea state, e.g., Fig. 3.10a–c, the numerical simulation shows a good agreement with the experiment again. The nonlinear effect can be ignored since the initial wave steepness is smaller than 0.11 in the first group. As anticipated, the third-order nonlinear MER model is almost reduced to Rayleigh distribution.

As for the moderate sea state, e.g., Fig. 3.10d–f, the NLS equation still performs the simulation reasonably well except for the initial stage where the wave height in the experiment is Rayleigh distributed. Still, the numerical result has presented a significant divergence in the tail. The reason for such a discrepancy is normally complex at this stage. The main one is due to the earlier developed nonlinearity in the numerical simulation, and the other one can be attributed to the uncertainty and variability in statistics.

As the wave propagates downstream in the basin, modulational instability will significantly work on the evolutionary process, as reflected by the perfect fit of the third-order MER model to the larger wave height distribution. The number of waves in the time series also plays a significant role in predicting extreme wave height, particularly in the long-term evaluation.

In the most severe sea state, e.g., Fig. 3.10g–i, the same conclusion can be drawn that the NLS equation still captures the main characteristics of extreme wave heights and that the agreement and disagreement can be explained in the same reason as before.

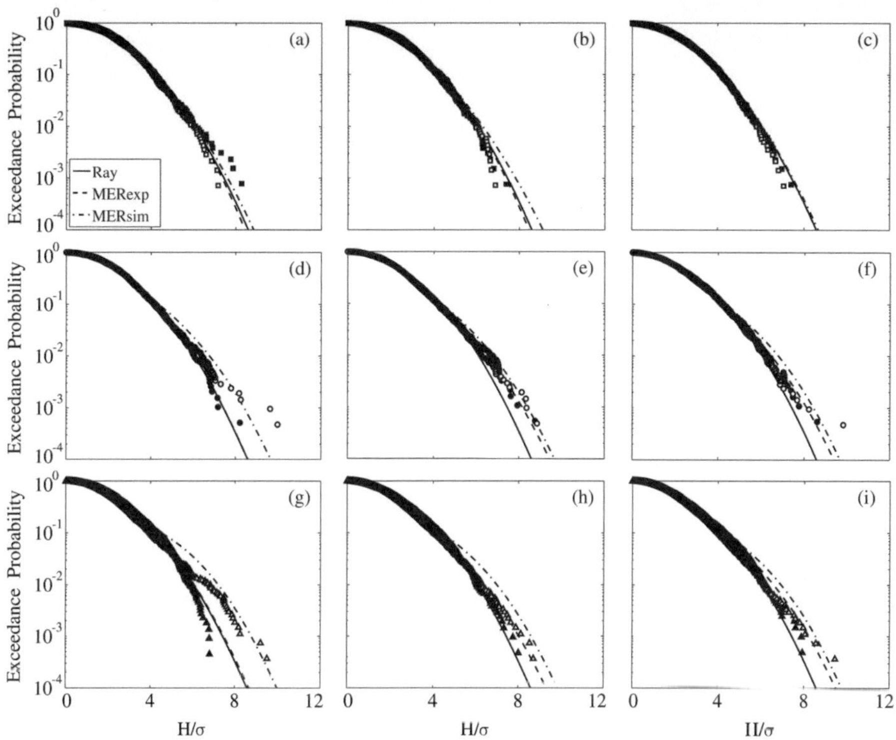

Fig. 3.10 Exceedance distributions of scaled wave heights in the CEHIPAR dataset. The three rows correspond to Case 4, 13, and 23 in Table 2.4. The three columns present the results obtained from Gauges at 20 m, 80 m and 120 m away from the wavemaker in sequence. The solid line, dash line and dot-dash line mean Rayleigh distribution, MER distributions in experiments and simulations, respectively. The full and empty marks still have the same meaning as before

Based on the above discussions in Figs. 3.8 and 3.10, it seems that in the presence of strong nonlinearity, a significant uncertainty can be observed between numerical simulations and laboratory observations, especially in the change process from linear to fully developed nonlinear states. Therefore, it makes sense to attempt to use relative rather than absolute distance to evaluate the evolutionary property in the following analysis.

It is also detected that the numerical simulation presents larger wave heights than the experiments, which is attributed in part to the energy dissipation, such as wave breaking in reality (Bitner-Gregersen and Toffoli 2012). In other words, the influence of wave breaking must be considered to give an exact description of the tail of the exceedance probability.

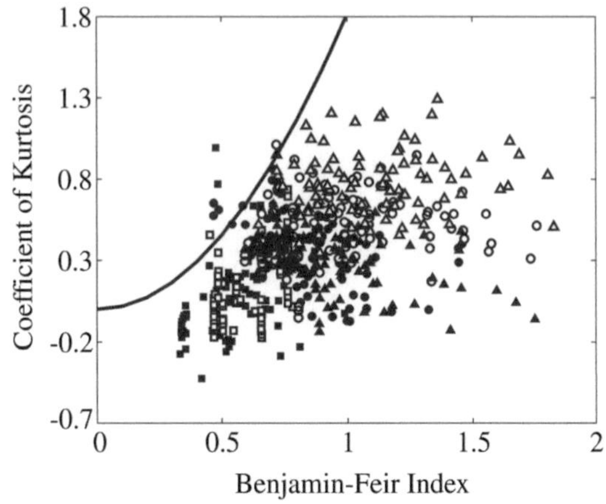

Fig. 3.11 Coefficient of kurtosis versus *BFI* in the CEHIPAR dataset

3.4.2 Statistics on Maximum Wave Heights

The relationship between the coefficient of kurtosis and *BFI* is depicted in Fig. 3.11 for the CEHIPAR wave basin. In conformity with the results presented in Fig. 3.3, the quadratic equation (solid line) represented by Eq. (3.4) does overestimate the results of experiments and simulations. Comparing with the numerical simulations (empty triangle), it can be concluded that wave breaking is severe in laboratory experiments (full triangle) in high sea states. As stated before, it obviously generates a smaller kurtosis coefficient value and consequently leads to a further deviation from the theoretical prediction.

In Fig. 3.12, a similar linear relationship as that illustrated in Fig. 3.4 can also be detected in the CEHIPAR dataset. Thus, it is confirmed again that the behaviour of some statistical quantities such as λ_{40} is consistent with that of extreme events in the time series (Guedes Soares et al. 2004). In other words, the coefficient of kurtosis can be considered the representative statistical parameter concerning the probability of occurrence of abnormal waves (Hagen 2002). Meanwhile, it is observed again that the simulated results are sometimes larger than those in the experiments, as clearly explained in Sect. 3.3.1.

As a result of the interest in understanding rogue wave generation, the maximum achievable wave height during the evolution of an unstable wave train has been investigated in the past. The first study is conducted in the wave flume, aimg to describe the steepness of the maximum wave as a function of the initial steepness of the wave train. Later, the results are further compared with the cubic NLS solution, showing that the simulation tends to generate much higher maximum wave steepness than the tank experiment. Compared with another tank experiment carried out in Tokyo by Waseda (2005), a systematic deviation from the results can be observed with a higher value due to the controlled perturbations.

In the experiments, the maximum wave heights are computed from zero up-crossing and down-crossing waves, respectively. Thus, four scaled maximum wave heights are

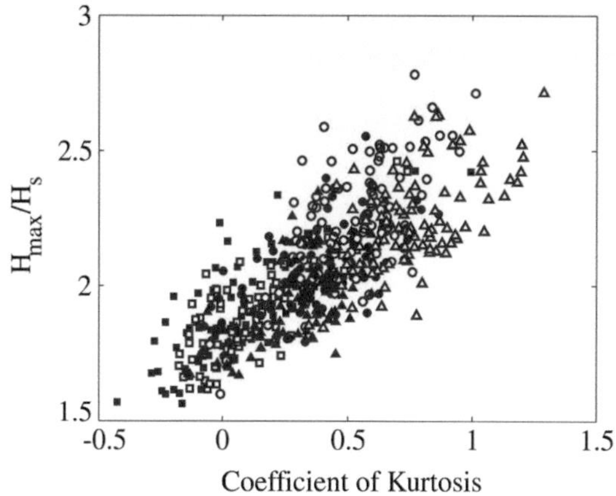

Fig. 3.12 Scaled maximum wave height versus coefficient of kurtosis in the CEHIPAR dataset

derived from the two series for each sea state, and their mean value is displayed in Fig. 3.13. It is very interesting that the largest scaled wave height appears in the moderate sea state, where the initial steepness $\varepsilon \approx 0.14$ rather than in the severe sea state. This variation is consistent with the previous physical experiment. The numerical simulation also presents the same tendency, although a certain degree of overestimation is observed in moderate and severe sea states. Considering that there is no energy dissipation such as wave breaking in the numerical simulation, this change should be partly attributed to the complicated nonlinear effects.

The same procedure for processing wave data is adopted in Fig. 3.14 where the relationship between the steepness of maximum wave height and the initial Benjamin–Feir Index is manifested (Ponce de Leon and Guedes Soares 2022). In the experiments, one similar variation to Waseda's results (2005) appears despite the difference in wave systems (continuous spectrum vs. three harmonic waves). The numerical results are also consistent

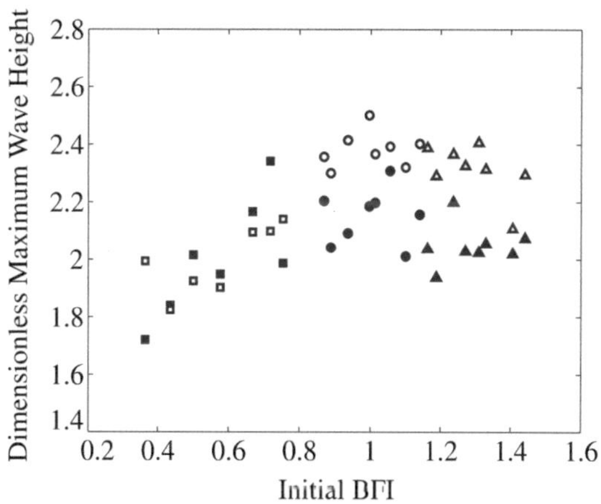

Fig. 3.13 Dimensionless maximum wave height versus initial *BFI*

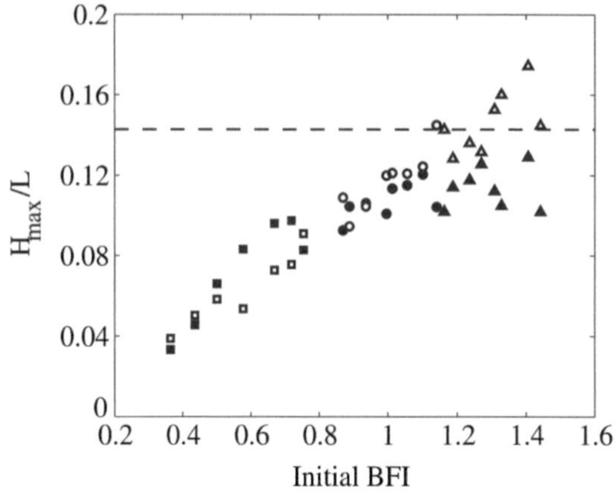

Fig. 3.14 Steepness of maximum wave height versus initial *BFI*

with those derived from numerical simulations of higher-order nonlinear equations, i.e., Dysthe and Zakharov equations (Waseda 2005).

Apparently, in Fig. 3.14, a good agreement between simulation and experiment can be detected except for the severe sea state, and this discrepancy is mainly due to the energy dissipation in the form of wave breaking. Moreover, many observations have proved that the maximum wave steepness, known as Miche–Stokes limit, which is close to 0.1429 in deep water and depicted by the dashed line in Fig. 3.14, could be exceeded (Toffoli et al. 2010b). Still, the mean steepness of maximum wave height in the laboratory experiments does not exceed the Stokes limit. Thus, it reveals that the wave shape change, in that case, is more related to the reduction of the wavelength rather than the increase of wave height (Toffoli et al. 2010b).

3.5 Conclusion

Generally speaking, the speed and type of the evolution of a wave packet governed by the NLS equation strongly depend on the initial conditions. The envelope pulse will eventually disintegrate into several envelope pulses or solitons that are stable enough to propagate. In the absence of dissipative effects, the end state of the evolution of a nonlinear wave train in deep water is neither random nor steady but is a series of periodically recurring processes (FPU recurrence).

In the numerical simulation of the NLS equation, the discrepancy between Λ and Λ_{app} cannot be exactly grasped, even including the second-order bound wave effect while in laboratory experiments, except for the smooth sea state, Λ_{app} is always a little larger than Λ and the difference will be slightly enlarged with the increased initial steepness.

Removing the effects of bound waves, Λ is almost equal to Λ_{app} in all sea states in CEHIPAR. Consequently, the third-order GC and MER models make no significant difference in predicting extreme wave heights. Generally speaking, the MER or GC model tends to the linear Rayleigh distribution due to the insignificant nonlinearity in the low sea state; as the evolution of waves approaches the fully developed condition in the moderate sea state, the third-order MER or GC model can describe the larger wave heights reasonably well; in the severe sea state, MER or GC model still works but is strongly affected by the severe wave breaking in the experiment.

As expected, the relationship between the coefficient of kurtosis and *BFI* is overestimated by the quadratic function because the analytical expression is derived for the nonlinear steady state at infinity while the experimental results are from all wave gauges in the wave basin. The scaled maximum wave height presents a highly correlated linear relationship with the coefficient of kurtosis. Thus, to a certain degree, λ_{40} can indicate the presence of extreme events in the time series. In addition, the extreme wave height also depends on the number of recorded waves.

Laboratory experiments reveal that the maximal scaled wave height appears in the moderate sea state where the initial steepness $\varepsilon \approx 0.14$ rather than in the severe sea state, and the numerical simulations also present a similar variation tendency. Considering that there is no energy dissipation in the numerical computation, this phenomenon should be partly attributed to the complicated nonlinear effect. For the severe sea state represented by a larger initial *BFI*, the steepness of the maximum wave is usually large. However, the laboratory experiment does not exceed the Stokes limit, while the exceedance has been reported in many published works. Thus, it indicates that extreme steepness is more likely due to the reduction of wavelength than an increase in wave height.

4 Modelling Extreme Waves with 2D Dysthe Equation

4.1 Basic Theory

4.1.1 2D Dysthe Equation

For the inviscid, irrotational and incompressible water waves propagating in horizontal directions, the envelope equations can be obtained by employing a harmonic expansion of the velocity potential ϕ and the surface displacement η under the hypothesis of small amplitudes and quasi-monochromatic waves (Mei, 1989). Truncating the power series expansion to include the third-order nonlinearity, the NLS equation can be derived while further operation, the Dysthe equation can be recovered.

Dysthe equation also called the modified nonlinear Schrödinger equation (MNLS), considers the higher-order terms responsible for nonlinear dispersion, full dispersion of linear waves and the effect of induced long-scale current.

Trulsen et al. (2000) explained how the MNLS equation can be enhanced with the exact linear dispersion by introducing a pseudo-differential operator for the linear part. By expanding the linear pseudo-differential operators in power series expansions and truncating at appropriate orders, the modified nonlinear Schrödinger equation of Dysthe (1979) and the broader bandwidth equation of Trulsen and Dysthe (1996) can be obtained, respectively. Furthermore, by truncating the nonlinear part to retain only the leading cubic nonlinear term, the standard nonlinear Schrödinger equation can be derived as well. In one horizontal dimension, the operator is algebraic, leading to the following MNLS equation with a moving frame at the wave group velocity:

$$\underbrace{\frac{\partial A}{\partial x}}_{\text{I}} + \underbrace{i\frac{k_0}{\omega_0^2}\frac{\partial^2 A}{\partial t^2}}_{\text{II}} + \underbrace{ik_0^3|A|^2A}_{\text{III}} - \underbrace{\frac{8k_0^3}{\omega_0}|A|^2\frac{\partial A}{\partial t}}_{\text{IV}} - \underbrace{\frac{2k_0^3}{\omega_0}A^2\frac{\partial A^*}{\partial t}}_{\text{V}} - \underbrace{i\frac{4k_0^3}{\omega_0^2}A\left.\frac{\partial \bar{\phi}}{\partial t}\right|_{z=0}}_{\text{VI}} = 0, \tag{4.1}$$

H. Zhang and C. Guedes Soares, *Numerical Modelling of Extreme Waves*, Synthesis Lectures on Ocean Systems Engineering, https://doi.org/10.1007/978-3-031-77084-5_4

$$\frac{\partial \overline{\phi}}{\partial z} = -k_0 \frac{\partial |A|^2}{\partial t}, \qquad z = 0, \tag{4.2}$$

$$\frac{\partial^2 \overline{\phi}}{\partial z^2} + 4\frac{k_0^2}{\omega_0^2}\frac{\partial^2 \overline{\phi}}{\partial t^2} = 0, \qquad -\infty < z < 0, \tag{4.3}$$

$$\frac{\partial \overline{\phi}}{\partial z} = 0, \qquad z = -\infty. \tag{4.4}$$

In Eq. (4.1), $\overline{\phi}(x, z, t)$ is the first approximation for the potential of the induced mean current. Note that the first three terms of Eq. (4.1) constitute the NLS equation, while the remaining parts, IV, V, and VI are related to the higher-order nonlinear dispersive effects that account for dynamics characterized by a broader spectrum with respect to the NLS equation. The last term VI is the result of the coupling with the mean flow, i.e. the potential associated with very long waves that result from three-wave non-resonant interactions, while the other higher-order terms are related to the self-steepening of the envelope. The complex amplitude function $A(x, t)$ and the surface elevation $\eta(x, t)$ are linked by the following relation:

$$\eta(x,t) = -\frac{k_0}{\omega_0^2}\overline{\phi}_t + \mathrm{Re}\Bigg\{A\exp[i(k_0x-\omega_0t)] + \left(\frac{k_0}{2}A^2 + \frac{ik_0}{\omega_0}A\frac{\partial A}{\partial t}\right) \times \exp[2i(k_0x-\omega_0t)] + \frac{3k_0^2}{8}A^3\exp[3i(k_0x-\omega_0t)]\Bigg\}. \tag{4.5}$$

The first term in Eq. (4.5) represents the induced long-wave component, the second term constitutes the contribution of free waves, and the other terms are phase-locked to the complex envelope $A(x, t)$, representing the contribution of the second and third bound wave harmonics.

4.1.2 Statistical Models

In the frame of linear wave theory, the propagation of irregular waves can be represented by the sum of a large number of harmonic and statistically independent wave components. Thus, according to the central limit theorem, as the number of harmonics increases infinitely, the wave surface displacement at one point will tend to Gaussian distribution:

$$f(\eta) = \frac{1}{\sqrt{2\pi}\sigma}\exp\left[-\frac{\eta^2}{2\sigma^2}\right], \tag{4.6}$$

where σ is the standard deviation of surface elevation η.

If the wave spectrum is strictly narrow-banded, the envelopes of surface elevations will vary such that the maxima and minima of the upper and lower envelopes coincide exactly with crests and troughs. Under this circumstance, the crest and trough will present the same Rayleigh exceedance distribution:

$$E_R(z) = \exp\left(-\frac{1}{2}z^2\right), \tag{4.7}$$

where the crest or trough has been scaled with σ.

In nonlinear waves, the second-order harmonics introduce a vertical asymmetry to the surface profile, rendering wave crests sharper and higher and troughs shallower and more rounded. Hence, the probability density function of the surface elevation has to be further modified. The approximated form of Tayfun second-order distribution was given by Socquet-Juglard et al. (2005):

$$f(\eta) = \frac{1 - 7\sigma^2 k_p^2 / 8}{\sqrt{2\sigma^2\pi\left(1 + 3G + 2G^2\right)}} \exp\left(-\frac{G^2}{2\sigma^2 k_p^2}\right), \tag{4.8}$$

where

$$G = \sqrt{1 + 2k_p\eta} - 1, \tag{4.9}$$

and k_p corresponds to the peak frequency by the linear deepwater dispersion relationship.

In addition, the exceedance distributions of crest ξ^+ and trough ξ^- amplitudes both deviate noticeably from the Rayleigh form of Eq. (4.7) and approach the following asymptotic distributions, valid for relatively large waves in the most general cases (Tayfun and Fedele, 2007; Fedele and Tayfun, 2009).

$$E_{\xi^+}(z) = \exp\left[-\frac{1}{2\mu^2}\left(-1 + \sqrt{1 + 2\mu z}\right)^2\right], \tag{4.10}$$

$$E_{\xi^-}(z) = \exp\left[-\frac{1}{2}z^2\left(1 + \frac{1}{2}\mu z\right)^2\right], \tag{4.11}$$

where μ represents a dimensionless measure of wave steepness and can be defined in various forms (Tayfun, 2006).

For unidirectional waves, taking into account the third-order nonlinearity, the preceding expressions are further modified as

$$E_{\xi^+}(z) = \exp\left[-\frac{1}{2\mu^2}\left(-1 + \sqrt{1 + 2\mu z}\right)^2\right]\left[1 + \frac{\Lambda}{64}z^2\left(z^2 - 4\right)\right], \tag{4.12}$$

$$E_{\xi^-}(z) = \exp\left[-\frac{1}{2}z^2\left(1 + \frac{1}{2}\mu z\right)^2\right]\left[1 + \frac{\Lambda}{64}z^2\left(z^2 - 4\right)\right], \tag{4.13}$$

where Λ has the same expression as that given in Eq. (3.5).

4.2 Effects of Fourth-Order Terms in Dysthe Equation

For a better understanding of the capability of Dysthe equation in describing the evolution of propagating wave packets, one simple but typical wave envelope adopted before will still be used here with the initial carrier wave period $T_0 = 2\pi / \omega_0 = 7$ s and forcing amplitude $a_0 = 1.75$ m.

This section can be considered as a continuity of research in Sect. 3.2 (Shemer et al. 2002; Zhang et al. 2014). The evolution of a deterministic wave group is numerically investigated in deepwater conditions with a series of linear and nonlinear models, of which the NLS and Dysthe equations can reproduce the Benjamin–Feir instability.

As depicted in Fig. 4.1, all initial wave envelopes in numerical simulations display the same Gaussian shape, derived by Hilbert transform of the surface elevation expressed in Eq. (3.12) the produced wave envelopes in the evolutionary process are all normalized by the initial forcing amplitude.

The left panel of Fig. 4.2 shows that the spatial variations of the same initial Gaussian envelope exhibit a different evolutionary tendency under different governing equations. When the evolutionary process is linear, there is no energy focusing, and the dispersive effect dominates the propagation of the wave envelope, as presented in Fig. 4.2a.

Once the third-order nonlinear effect is included, the simulation, governed by the NLS model, not only keeps the symmetry of the initial wave packet in the whole process but also adjusts the shape and width of the envelope to become a fundamental soliton with oscillatory tails, attaining a "sech" profile. It seems in this case, i.e. Figure 4.2b, two smaller solitons only appear in the space from $20L_0$ to $28L_0$ and then merge with the leading soliton once again, accompanied with less energy spread to the front and back of the wave packet due to the presence of a third-order nonlinear effect.

If the higher-order nonlinearity is added, as exhibited in Fig. 4.2c, the simulation controlled by Dysthe model now can exactly reproduce the asymmetric property of

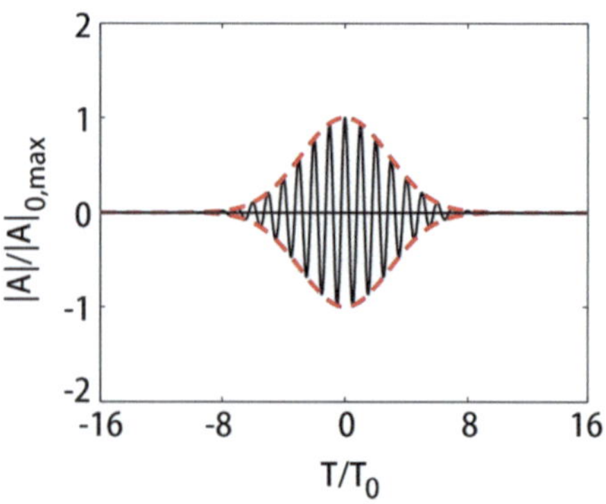

Fig. 4.1 Initial wave envelope in all numerical simulations

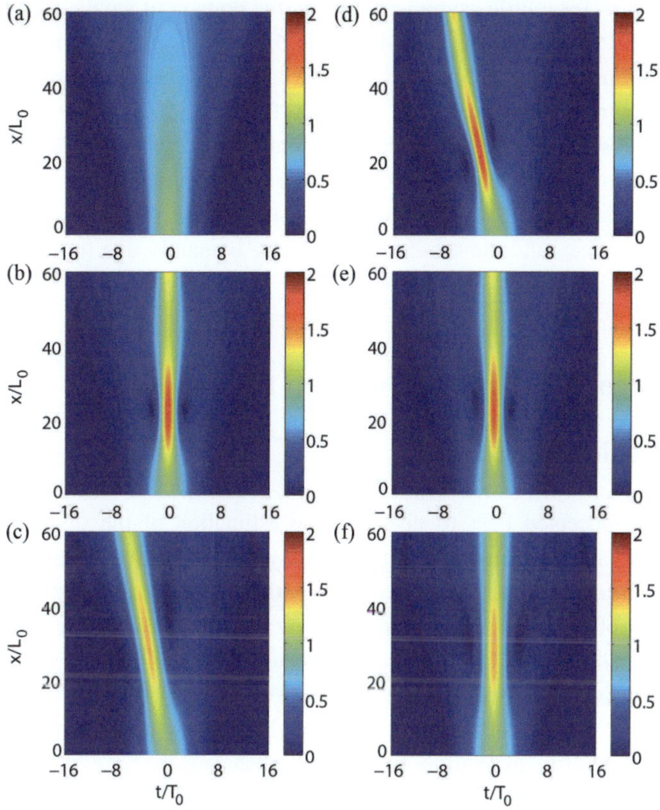

Fig. 4.2 Spatial variations of wave envelopes: **a** Linear model; **b** NLS model; **c** Dysthe model; **d** NLS + term IV; **e** NLS + term V; **f** NLS + term VI

wave packet (Shemer et al. 2010a). The Gaussian wave packet splits into two soliton-like envelopes asymmetrically, with the main energy part moving in an increased group velocity, which is in qualitative agreement with the observation of Feir.

Moreover, the energy-focusing process in the Dysthe equation is much weaker and slower than that in the NLS equation, meaning that the extreme wave height is lower and rogue waves appear over a longer distance.

To indicate the effect of each fourth-order term on the solution of Dysthe model, one approach has been adopted herein, originally from the work of Shemer et al. (2002), where the profiles of Gaussian packets are only compared in three spatial locations.

In Fig. 4.2d, the governing equation is composed of the NLS model and term IV in Eq. (4.1). This term mainly leads to the visible asymmetry of envelope shapes and reinforces the nonlinear focus effect. Compared with Fig. 4.2b, there is no large difference in Fig. 4.2e except for a little asymmetry around the domain of smaller solitons, meaning the term V in Eq. (4.1) has little contribution to the higher-order nonlinear effect.

The function of the induced mean current is also considered by the addition of term VI into the NLS model. As presented in Fig. 4.2f, this modification does not violate the symmetry of the solution but leads to the widening of the envelope shape and suppresses the maximum wave amplitude compared with the NLS solution in Fig. 4.2b.

Thus, it can be inferred that the induced mean current plays a negative role on the nonlinear focus effect, and the envelope asymmetry is mainly due to the contribution of term IV.

4.3 Comparison Between NLS and Dysthe Models

4.3.1 Statistical Moments

It is commonly accepted that, to give a stable estimation of the third-order nonlinear effect induced by modulational instability, a large number of wave series is mandatory. To eliminate the influence of statistical variability in the comparison of statistical parameters, an ensemble of 50 time simulations governed by the NLS and Dysthe models has been constructed for all cases in Table 2.2. To indicate the variability and uncertainty of the statistical estimation, the error bars are also plotted in Fig. 4.3.

It needs to be pointed out that the numerical simulations performed by NLS equation in this chapter are different from those in Chap. 3 because the second-order bound wave effect is included now.

For rogue wave density, the error bars are computed as $\pm p_{rog}/\sqrt{N_{rog}}$ where p_{rog} is the obtained rogue wave density that has the same definition as Sect. 3.3.1, i.e., the ratio of the number of wave heights N_{rog}, including both zero up-crossing and down-crossing waves which are larger than double of the significant wave height, to the total number of recorded wave heights.

In wave statistics, surface elevation skewness and kurtosis coefficients usually express the higher-order statistical moments related to wave nonlinearity. As shown in Fig. 4.3a, the two numerical simulations are almost identical in the coefficient of skewness, although both exhibit slightly larger values than the experimental results. In other words, the vertical asymmetry provoked by the bound wave effects makes no significant difference between the NLS and Dysthe models.

As for the coefficient of kurtosis, λ_{40}, its evolutionary tendency downstream along the wave tank is plotted in Fig. 4.3b, accompanied by the rogue wave density in Fig. 4.3c. It is distinct that the evolutionary tendency of the coefficient of kurtosis is almost in agreement with that of rogue wave density in both numerical simulations, being able to catch the variation trend of laboratory experiments.

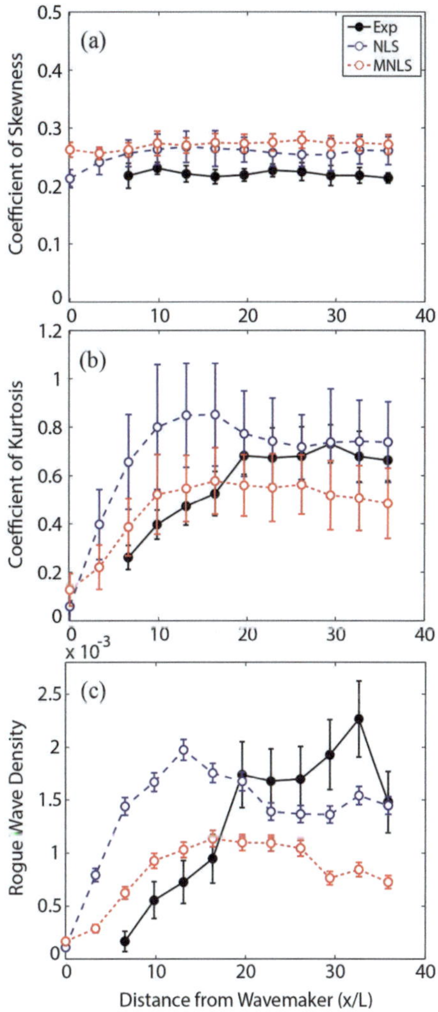

Fig. 4.3 Comparison of spatial variations of wave statistical parameters between numerical simulations and laboratory experiments in the case of 8241–45. **a** coefficient of skewness, **b** coefficient of kurtosis, **c** rogue wave density

Moreover, compared to the Dysthe model, the simulation of the NLS equation presents a more substantial nonlinear effect, demonstrated by the earlier achieved fully developed condition, which conforms with the former conclusions in the analysis of Gaussian wave packet in Sect. 4.2. Therefore, the numerical simulation performed by NLS equation will give a larger coefficient of kurtosis and, consequently a larger rogue wave density.

Finally, as indicated by the big error bars in Fig. 4.3b, the kurtosis coefficient variability is very large even though a vast number of data have been used. Thus, a large discrepancy cannot be avoided even though the same experiments are repeated many times. Concerning the rogue wave density, a large amount of series can reduce the statistical uncertainty,

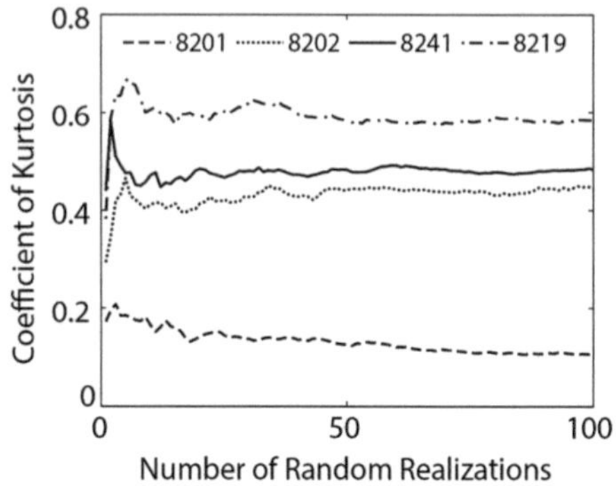

Fig. 4.4 Coefficient of kurtosis as a function of the number of random realizations in the numerical simulation governed by Dysthe equation

as indicated in Fig. 4.3c. Hence, it can be concluded that the amplitude of abnormal waves rather than the number of rogue waves increases within a large set of wave series.

To further indicate the relationship between the coefficient of kurtosis and the number of simulated series, Fig. 4.4 shows some examples derived by Dysthe equation. It seems that five similar wave series with each 25 min is the minimum size of the ensemble required to clearly illustrate the fundamental characteristic of modulational instability, which is a quasi-resonant four-wave interaction process that takes place near the peak of the spectrum (Onorato et al. 2005; Mori et al. 2007).

What needs to be mentioned here is that different symbols are also used in this chapter to identify the results of different cases. Their meanings are a little different from the preceding chapter but still represent the smooth (8201, square), first moderate (8202, diamond), second moderate (8241–45, circle) and severe (8219, triangle) sea states.

In second-order narrow band wave theory, a quadratic function can be derived between the coefficients of kurtosis and skewness, and a similar relation was found by Guedes Soares et al. (2003) based on full-scale field data.

However, in Fig. 4.5, the quadratic relation only catches the tendency of variation without giving impressive agreement because the coefficient of kurtosis is notably influenced by the dynamics of the free modes of four-wave interactions, while the skewness is only weakly affected. To a certain degree, the stable relationship indicated by the numerical simulations is consistent with the experimental results in Marintek and those presented in the work of Mori et al. (2007).

As indicated in Fig. 4.6, even though the bound wave effects are included, Λ is still equal to Λ_{app} in the case of linear wave series, e.g., the sea state 8201 (full square). Once the nonlinearity increases, Λ can become a little smaller than Λ_{app}, which is opposite to the second-order theoretical conclusion obtained by Tayfun and Fedele (2007). According to the previous research (Zhang et al., 2014) in Chap. 3, modulational instability does not contribute much to the discrepancy between these two parameters without bound wave effects. Now, it seems that the combined effects of modulational instability and bound wave effects may cause the difference.

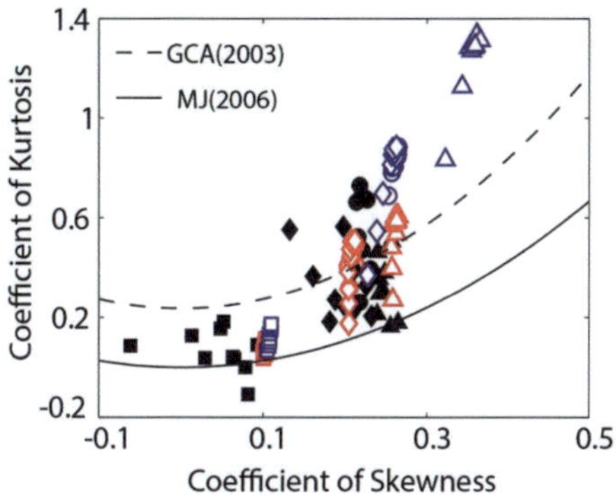

Fig. 4.5 Coefficient of kurtosis as a function of coefficient of skewness. The dashed line is the empirical fit by Guedes Soares et al. (2003), while the solid line is proposed by Mori and Janssen (2006). The black full symbols are experimental data, while the blue and red empty symbols are simulated results derived from NLS and MNLS equations. Different symbols represent different cases in Table 2.2

Fig. 4.6 Relationship between Λ and Λ_{app}. The meanings of marks are the same as those in Fig. 4.5

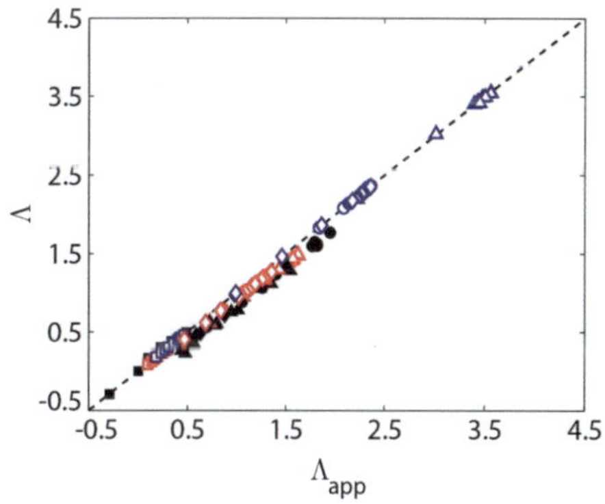

Furthermore, in the numerical simulation, the cubic nonlinear model overestimates the fourth-order cumulants in the severe sea state (8219) and cannot identify the difference between these two parameters even though including the second-order bound wave effect. On the other hand, the numerical model of Dysthe equation can provide a better description of their relationship, but it is still not good enough. The modulational instability enlarges the higher-order bound wave effects, which consequently leads to $\Lambda_{app} > \Lambda$. However, as common sense, a larger ensemble of nonlinear wave series usually leads to larger fourth-order cumulants, but this cannot be observed in the simulations of Dysthe equation.

According to the above discussion, the third-order nonlinear theoretical models associated with the parameter Λ will be adopted in the prediction of exceedance distribution of wave crest, trough and height.

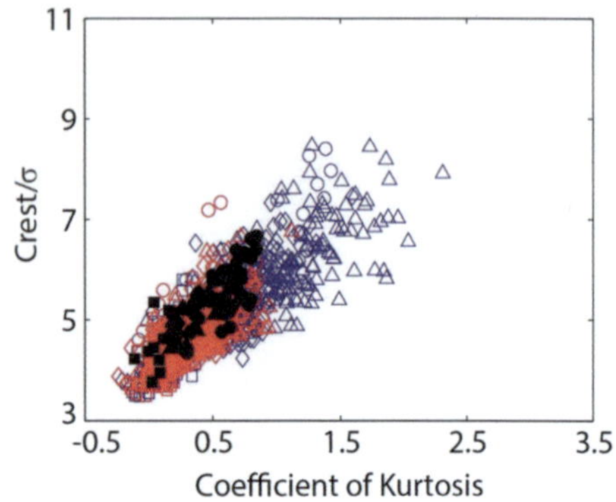

Fig. 4.7 Relationship between scaled maximum wave crest and coefficient of kurtosis. The meanings of symbols are the same as those in Fig. 4.5

It is well known that a linear regression model exists between the scaled maximum wave height and the coefficient of kurtosis, which is reported by many researchers, such as Guedes Soares et al. (2003), who analyzed all sea state records during the storm of November 1997 in North Alwin and Draupner storm at the beginning of 1995.

As indicated in Fig. 4.7, a similar linear relationship can be detected for the scaled maximum wave crests rather than for the scaled maximum wave troughs in both experimental and numerical results. As a result, it can be inferred that the abnormal wave is usually accompanied by a very large wave crest, apparently the formation of which is mainly attributed to the nonlinear interaction among free wave modes.

4.3.2 Surface Elevation

Figure 4.8 presents the probability density functions of the surface elevations measured at Gauge 8 where the modulational instability has almost been fully developed in all random sea states. For convenience, the surface elevations are scaled by the standard deviation σ of the concurrent time series. The experimental probability density function is also compared with the Gaussian distribution in Eq. (4.6) and the Tayfun second-order model prediction by Eq. (4.8).

As the initial wave steepness increases, i.e. Figure 4.8a–d, the wave surface elevations begin to deviate from the normal distribution noticeably, becoming narrower and more positively skewed that can be captured by the second-order model of Tayfun in most part except for the peak values in the severe (8219) and second moderate (8241) sea states. Therefore, the deviation in this range is mainly attributed to the bound wave effects. Meanwhile, the NLS-type simulations are consistent with the experimental results of up to 4 standard deviations in all sea states.

To further explore the nonlinear effect on the formation of larger surface elevation, the vertical axis has been changed to a logarithmic scale for the same distributions, as shown

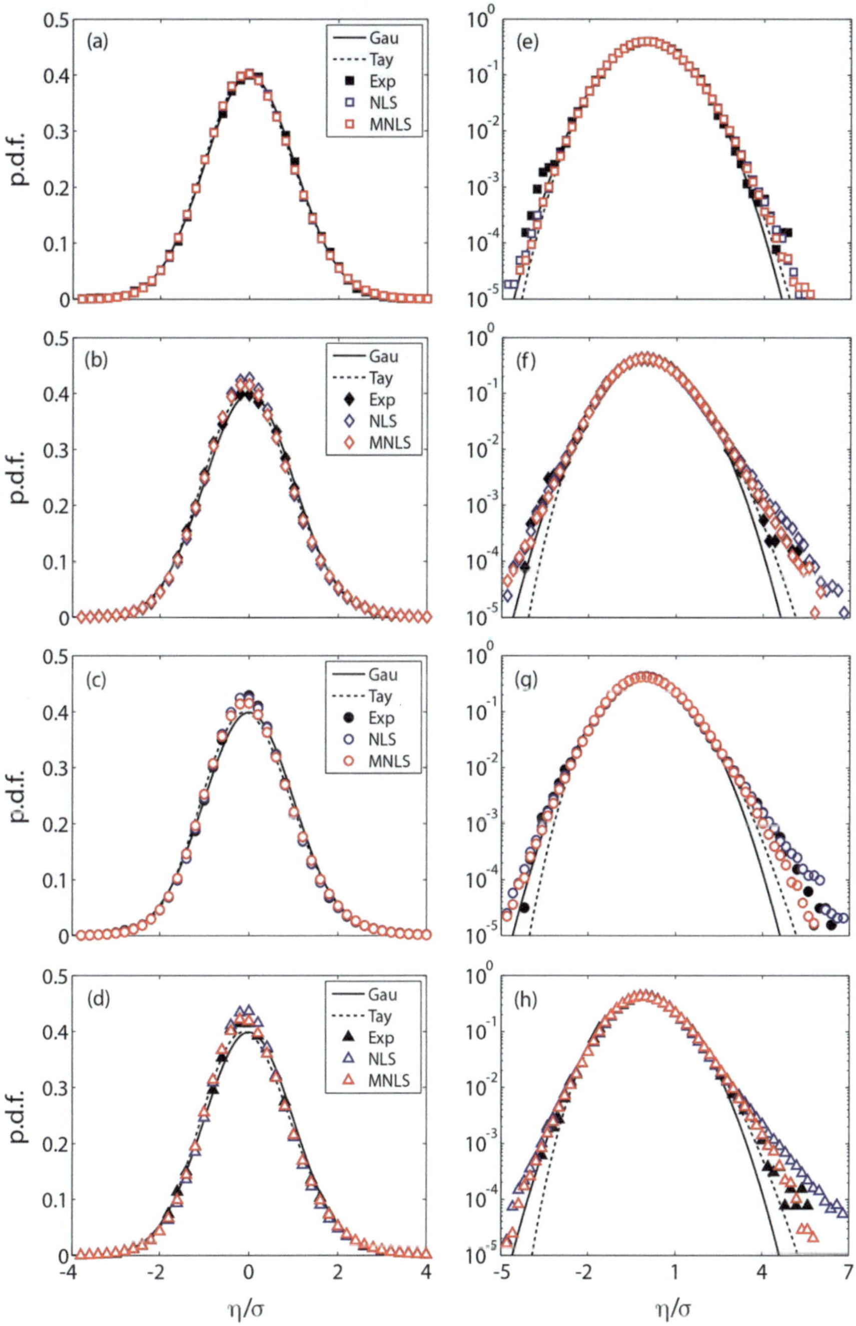

Fig. 4.8 **a–d** Probability density function of surface elevations in the location of Gauge 8 for different sea states. **e–h** are the same figures but on a logarithmic scale

on the right panel of Fig. 4.8. The positive skew indicates that the tail on the right side is longer and fatter than that on the left side.

In agreement with the previous experimental and numerical studies (Onorato et al. 2009; Socquet-Juglard et al. 2005; Toffoli et al. 2010a), if the wave series is strongly nonlinear, a substantial deviation from the Gaussian and second-order models can be observed on both tails of the wave surface distribution, in particular in the positive section. Considering the third-order or higher-order bound wave effect is much smaller, the considerable discrepancy in this part is mainly attributed to the modulational instability (Fedele et al. 2010).

Although the NLS-type numerical models are not fully nonlinear, they are both able to capture the aforementioned deviations reasonably well. Moreover, as the sea state becomes more and more severe, the NLS equation tends to overestimate the positive tail of wave surface distribution due to no energy dissipation and an unsuppressed stronger nonlinear effect (compared to Dysthe equation).

The spatial variation of surface elevation downstream along the wave basin is consistent with what was observed in the work of Onorato et al. (2009) where the directional spreading effect has been taken into account in various sea states.

4.3.3 Wave Height

The exceedance probability of wave heights is shown in Figs. 4.9 and 4.10. The comparison still focuses on the same locations as those analyzed in Chap. 3. To reduce the influence of irregularity induced by statistical uncertainty on the tail of wave height distribution, an ensemble of 18 time series has been constructed in numerical simulations. Meanwhile, the theoretical prediction of the linear Rayleigh model is presented as well.

In the low sea state, i.e., Figs. 4.9a–c, the initial Benjamin–Feir Index is equal to 0.5, much smaller than 1.0, meaning that the dispersive effect dominates the evolutionary process.

In this case, both NLS and Dysthe models can work well. The numerical results display a linear Rayleigh distribution all over the wave basin, which almost perfectly agrees with the experimental observations. However, a little discrepancy can be detected in Gauge 8, which could be attributed to the variability caused by only one wave series in the experiment.

At the initial stage of the first moderate sea state, i.e., Fig. 4.9d, the nonlinearity is not very strong; thus, both numerical simulations and laboratory experiments present a linear distribution. As the waves propagate downstream in the wave basin, modulational instability begins to work on the evolutionary process significantly in the numerical simulations. It is much more robust in the NLS model, as indicated by the more significant exceedance probability of the same extreme wave height in Fig. 4.9e.

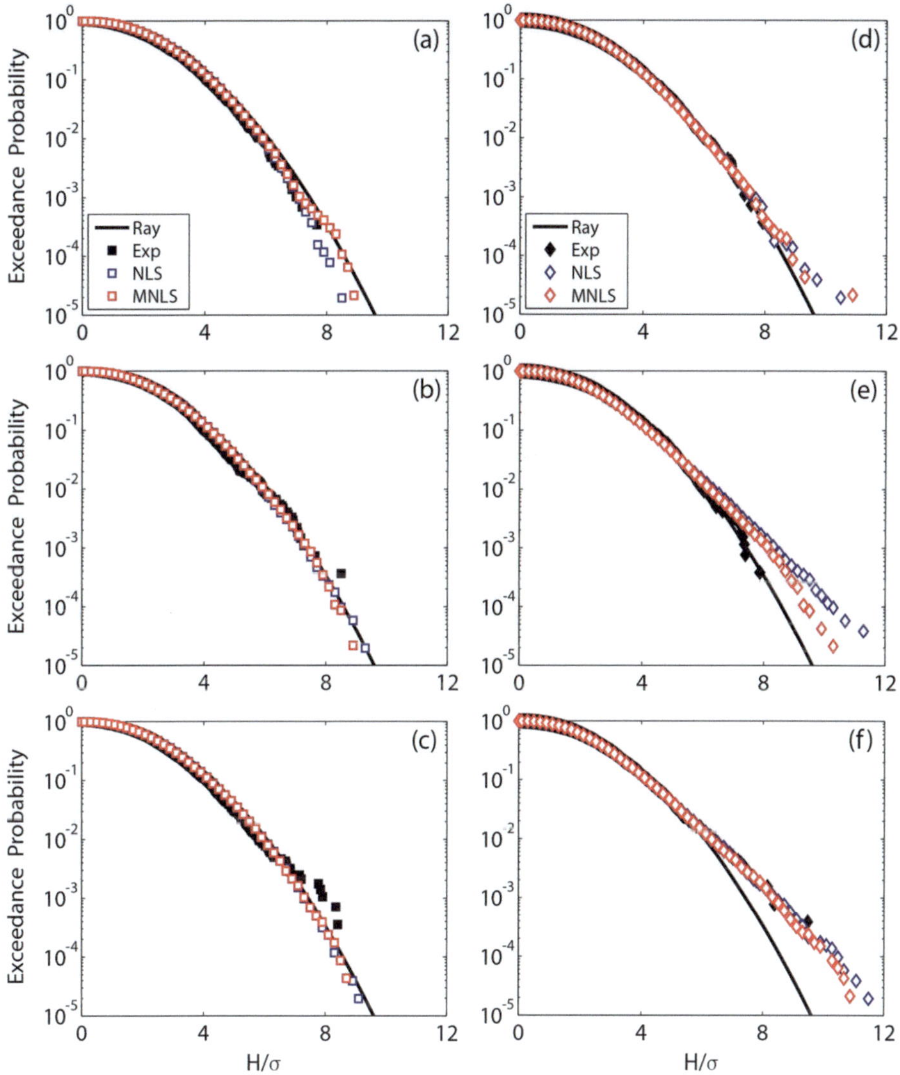

Fig. 4.9 Exceedance distributions of scaled wave heights in the low (8201) and first moderate (8202) sea states. **a–f** correspond to the cases 8201 and 8202 listed in Table 2.2, respectively. The three rows present the results obtained from Gauges 1, 5 and 8, located at the normalized distances $x/L = 3.2, 9.6, 14.4$, respectively

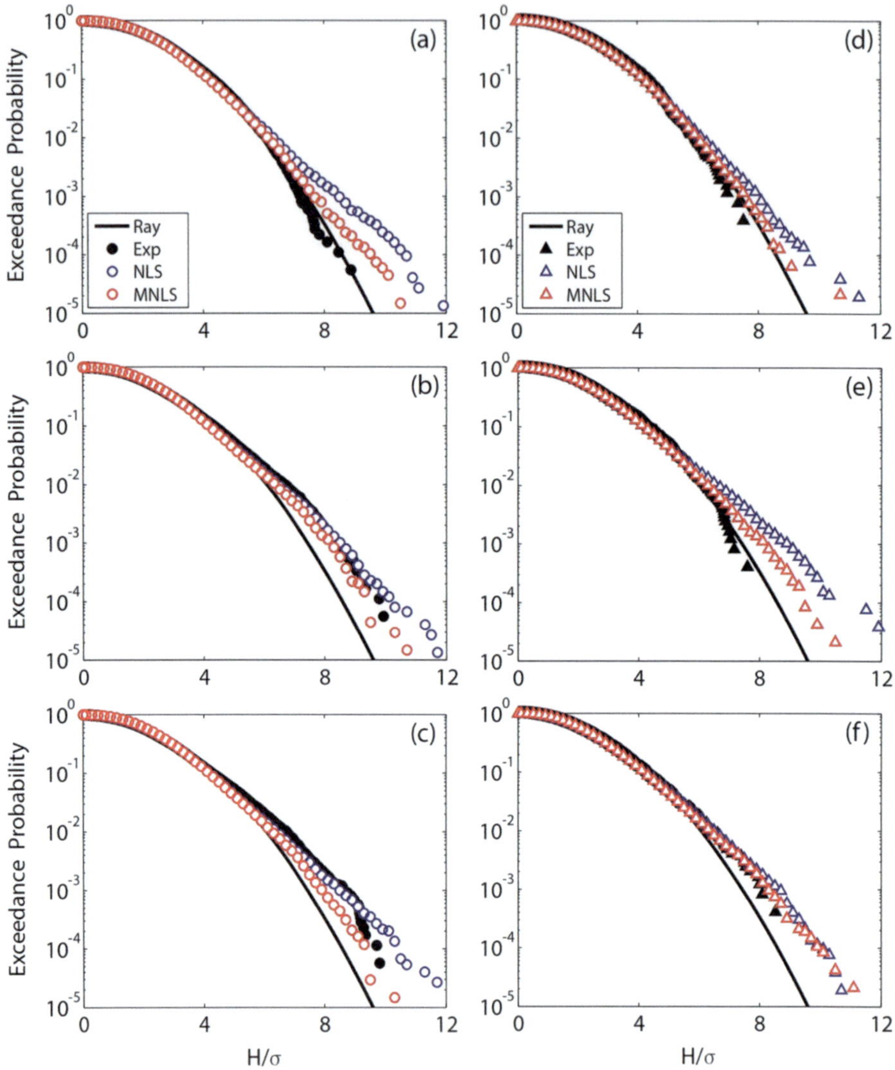

Fig. 4.10 Exceedance distributions of scaled wave heights in the moderate (8241–45) and severe (8219) sea states. **a**–**c** correspond to the case 8241–45 and the results at locations $x/L = 6.5$, 19.6, 29.4, respectively. **d**–**f** correspond to case 8219 and the same locations as those in Fig. 4.9

However, in laboratory experiments, the dispersive effect still dominates the evolutionary process, as indicated by the linear distribution of wave height. This divergence in evolutionary speed between numerical computations and laboratory experiments may be primarily attributed to the insignificant energy dissipation in simulations, which can accelerate the evolutionary process and provide a more intensive focus.

At the peak stage, not only numerical simulations but also laboratory experiments have reached a balanced state, revealed by the same nonlinear wave height distribution in Fig. 4.9f, which can be perfectly fit by the third-order nonlinear theoretical model (MER or GC) as argued in other work (Shemer and Sergeeva, 2009; Zhang et al. 2015).

In the second moderate sea state with a shorter dominant wave length L, it is possible to evaluate the much longer spatial evolution of nonlinear waves in the wave basin and consider the influence of the number of waves on estimating extreme wave height.

The difference in Fig. 4.10a and consistency in Fig. 4.10b between simulations and experiment can be explained in the same way as those in Fig. 4.9e and f since they are almost in the same scaled locations. As the waves continue to propagate, if the statistical uncertainty is ignored, the NLS simulation could still catch the main characteristic of extreme wave heights, while the Dysthe simulation gives a lower probability.

As for the severe sea state, the numerical simulations performed by NLS and Dysthe models can capture the main statistics of larger wave heights except for the intermediate stage in the wave evolutionary process, as illustrated in Fig. 4.10e, where the energy dissipation mainly due to the performance of wave breaking seriously influences the tail of wave height distribution and hence enlarges the deviation from numerical simulations.

4.3.4 Wave Crest

Figures 4.11 and 4.12 display the spatial variations of exceedance distribution of wave crests, defined as the highest elevation of each wave with respect to the mean water level. The second-order interaction should participate in the deviation from Gaussian statistics for the crest amplitude. Therefore, apart from the linear Rayleigh distribution of Eq. (4.7) and the third-order nonlinear model of Eq. (4.12), the second-order distribution in Eq. (4.10) is also compared with the experimental data.

In the low sea state (8201), at the initial and intermediate stages, i.e. Figure 4.11a, b, both second-order and third-order nonlinear theoretical models can describe the wave crest distributions very well because the nonlinearity is weak and arisen from the second-order bound wave effect. However, at the peak stage, i.e. Figure 4.11c, the third-order modulation begins to but does not fully work on the generation of a larger amplitude of crest, indicated by the apparent deviation of observed statistics from the second-order nonlinear model. The two kinds of numerical simulations are almost consistent with each other and fit the experiments perfectly from the point of view of statistics.

In the first moderate sea state (8202), Fig. 4.11d–f, only the third-order GC model can describe the crest-height statistics reasonably well in the whole evolutionary process, meaning that the bound wave effect is relatively weak compared with the Benjamin–Feir type modulational instability. Although the two numerical simulations begin to deviate from each other at the larger amplitude part, they both can catch the major statistics of wave crest in a short-term wave series.

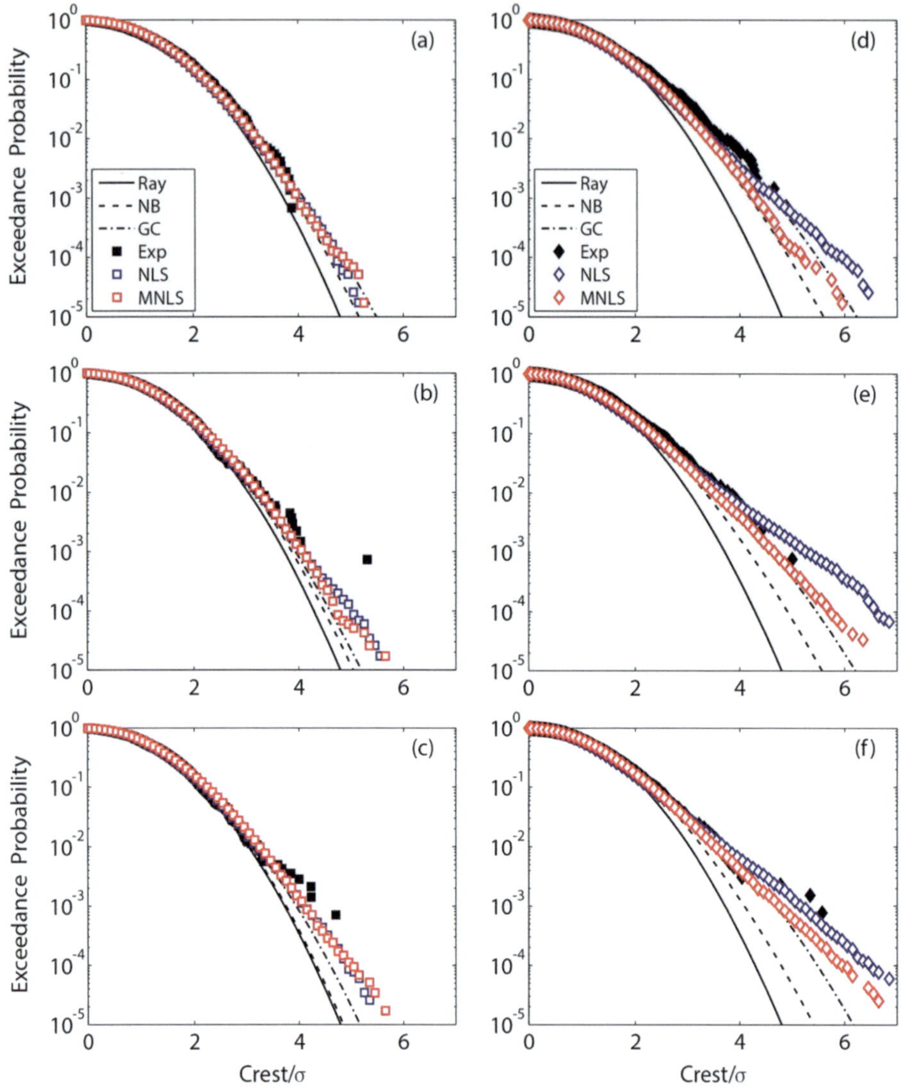

Fig. 4.11 Exceedance distribution of scaled wave crests in the low (8201) and first moderate (8202) sea states. **a**–**f** correspond to the first two cases listed in Table 2.2. The three rows present the results obtained from Gauges 1, 5 and 8, respectively. 'Ray', 'NB' and 'GC' represent linear, second-order and third-order models

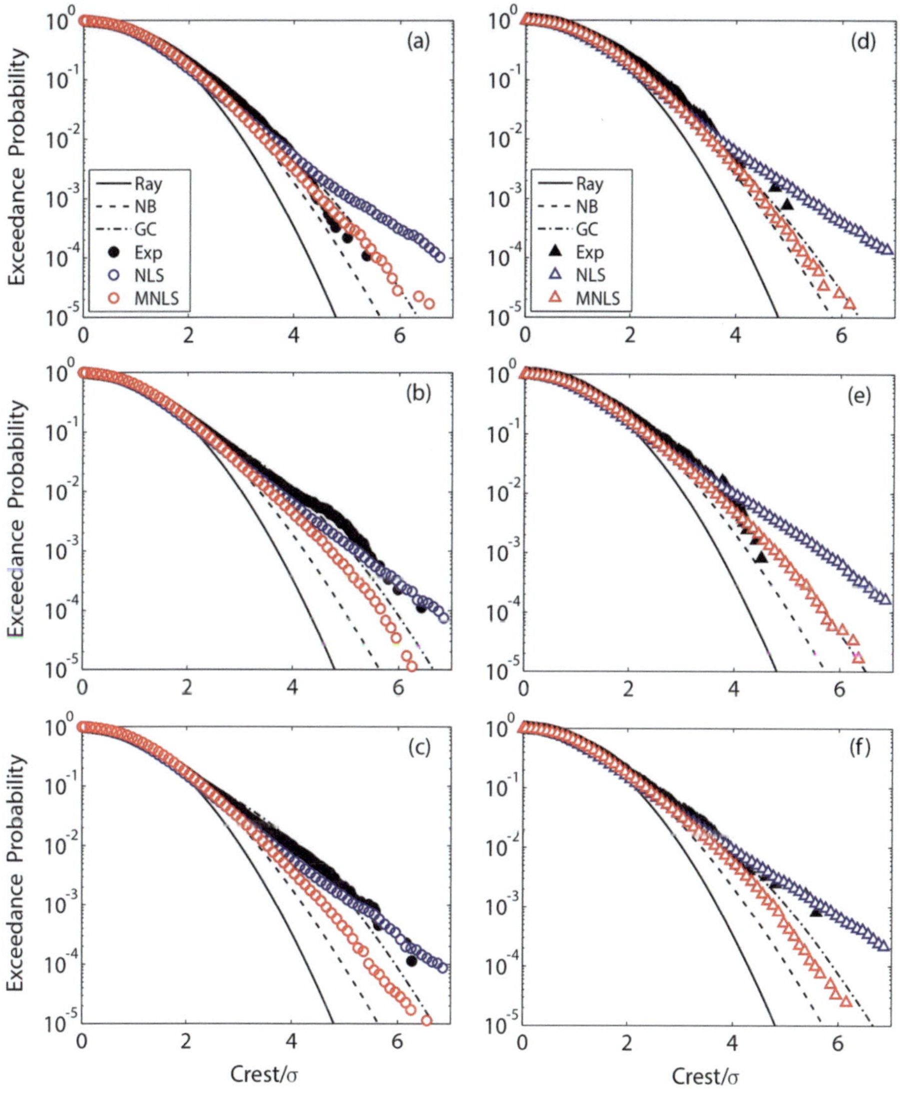

Fig. 4.12 Exceedance distribution of scaled wave crests in the second moderate (8241) and severe (8219) sea states. **a**–**f** correspond to the last two cases listed in Table 2.2. The three rows present the results obtained from Gauges 1, 5 and 8, respectively. ‘Ray’, ‘NB’ and ‘GC’ represent linear, second-order and third-order models

In the second moderate sea state (8241), five similar wave series with the same initial wave parameters not only guarantee a stable observation of the statistical variables such as the coefficient of kurtosis (Onorato et al., 2005), but also generate a much larger maximum crest height.

Additionally, the smaller wave peak period, in this case, means a much longer relative spatial evolution, up to 36 wave lengths corresponding to the peak of the spectrum at the wavemaker. It is found that the third-order GC model is still the best choice for describing the wave crest distribution. Moreover, as discussed on the extreme wave height distribution (Mori and Janssen, 2006; Mori et al., 2011), the number of waves also plays an important role in predicting extreme crest height. Besides, except for the initial stage, the numerical simulation of the NLS model is more consistent with the experimental results than that governed by the Dysthe equation.

In the severe sea state (8219), Figs. 4.12d–f, the third-order GC model still works, but some difference is noticeable on the tail of the distribution because in addition to the strong nonlinearity, wave breaking (see Fig. 4.12e) begins to play an important role on the evolutionary process as well, which cannot be included in the existed potential theory (Bitner-Gregersen and Toffoli 2012). With regard to the numerical simulation, it seems that the MNLS equation presents an agreement with the experiment in the initial and intermediate stages, while the NLS equation is in the peak stage.

However, the difference between MNLS and NLS equations should come from the spectral width, and neither of them fails in the local strong nonlinearity. Although not given here, it would be expected that the spectrum will be broadest at the peak stage, and therefore, NLS would not be a better approximation than MNLS. Hence, the controversy here may be attributed to the uncertainty and variability of the rough laboratory experiments. Agreement on the limited statistical estimation alone cannot be a very sound evidence for the validity of NLS at the peak stage.

4.3.5 Wave Trough

The same comparisons as those in wave crest distribution have also been applied to the wave trough. In the low sea state (8201), i.e. Figure 4.13a–c, the spatial variation of trough depth is more or less the same as that observed in the crest height. The numerical results governed by NLS-type equations tend to be Rayleigh distributed, which is compared favourably with the laboratory observations in most cases.

In the first moderate sea state (8202), i.e. Figure 4.13d–f, the second-order and third-order nonlinear models do not exhibit too much difference in fitting the experimental results at the initial and intermediate stages. Only in the peak stage has the effect of modulational instability been fully developed, causing the observed statistics to deviate from the conventional second-order NB model (Toffoli et al. 2008a).

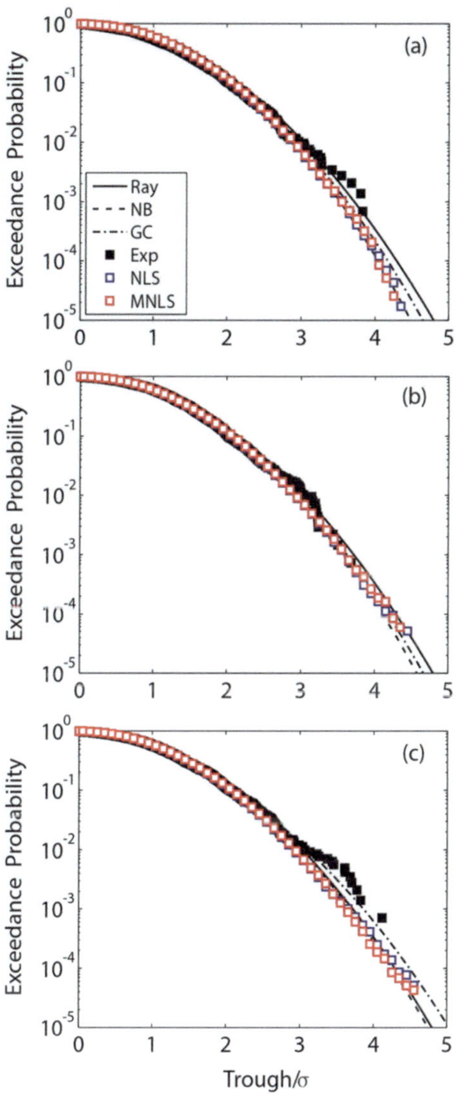

Fig. 4.13 Exceedance distribution of scaled wave troughs in the low (8201) and first moderate (8202) sea states. **a–f** correspond to the first two cases listed in Table 2.2. The three rows present the results obtained from Gauges 1, 5 and 8, respectively. 'Ray', 'NB' and 'GC' represent linear, second-order and third-order models

Compared with those deviations in wave crest under the same sea state, it can be concluded that wave trough is less sensitive than wave crest to modulational instability. Notably, the two numerical simulations are still in agreement with each other and fit the observed statistics reasonably well.

In the second moderate sea state (8241), the effect of modulational instability is not noticeable in Fig. 4.14a, meaning that wave trough evolves more slowly than wave crest. Considering that the coefficient of kurtosis also stops growing at Gauge 5, corresponding to the location of Fig. 4.14b, it can be argued that the third-order nonlinearity mainly contributes to the wave crest and that when crest and trough are both fully developed, the direct indication is the coefficient of kurtosis with an approximately stable value.

Moreover, as displayed in Fig. 4.14b, c, f, the wave trough usually can be fitted surprisingly well by the linear Rayleigh model if the wave series have been fully worked by the third-order four-wave interactions (Toffoli et al. 2008b; Shemer and Sergeeva, 2009).

As expected, wave breaking also influences the wave trough distribution, as illustrated on the tail of Fig. 4.14e, which is in agreement with the observed statistics of the crest in Fig. 4.12e. Although the NLS-type equations cannot include the effect of wave breaking, they have caught most observed statistics of wave troughs very well and kept in conformity with each other even in the presence of stronger nonlinearity.

Consequently, it can be said that the major discrepancy of wave height distribution between numerical simulations of NLS and Dysthe equations arises principally from the difference in wave crest distribution. In other words, wave crest is more sensitive than wave trough to the third-order quasi-resonant four-wave interaction.

4.3.6 Wave Period

A comparison of observed wave period distribution with the theoretical model of Longuet-Higgins (Longuet-Higgins, 1983) has been presented in the former chapter. This section only focuses on the numerical and experimental results measured in Gauge 8 because the spatial variation of the wave period is not very distinct except for being a little narrower downstream along the wave tank.

In Fig. 4.15a–d, it is evident that the periods of individual waves in a wave train exhibit a distribution narrower than that of wave crests or troughs, and the spread lies mainly in the range from 0.5 to 1.5 times the mean wave period, which is a little narrower than the scope of ocean waves proposed by Goda (2000).

Moreover, as the sea states become more and more severe (from top to bottom), the observed wave period distribution will become much steeper and narrower. It is also noticeable that the observed statistics generate a hump in the interval (0.5, 1) in the low sea state (see Fig. 4.15a), and similar phenomena are also observed in the laboratory experiments carried out in the offshore basin of CEHIPAR, Spain (Zhang et al., 2014).

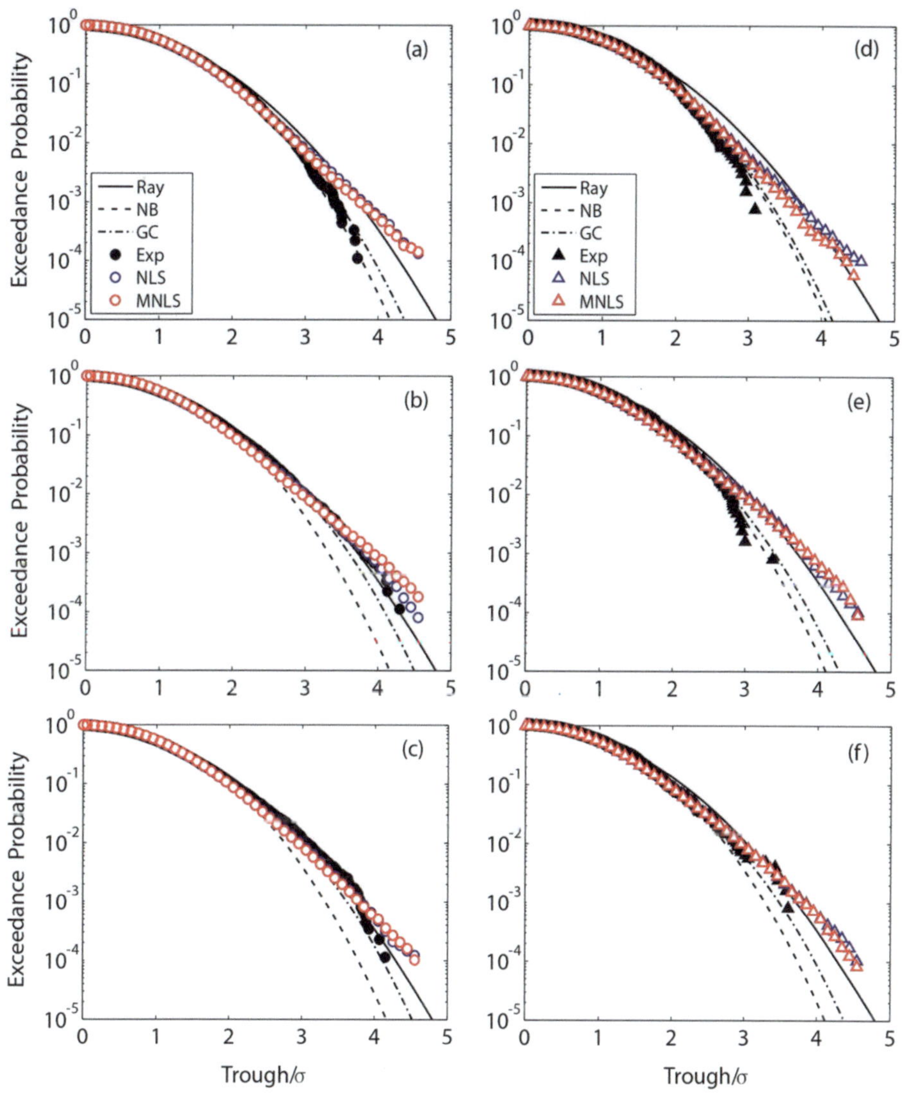

Fig. 4.14 Exceedance distribution of scaled wave troughs in the second moderate (8241) and severe (8219) sea states. **a–f** correspond to the last two cases listed in Table 2.2. The arrangement and the meanings of lines are the same as those in Fig. 4.13

The NLS-type numerical simulations can catch the fundamental properties of wave period distribution in all cases, particularly in the first moderate sea state (8202), i.e. Figure 4.15b. It is also surprising that the tails on the left and right sides of the wave period distribution fit quite well by both numerical simulations, which are presented in

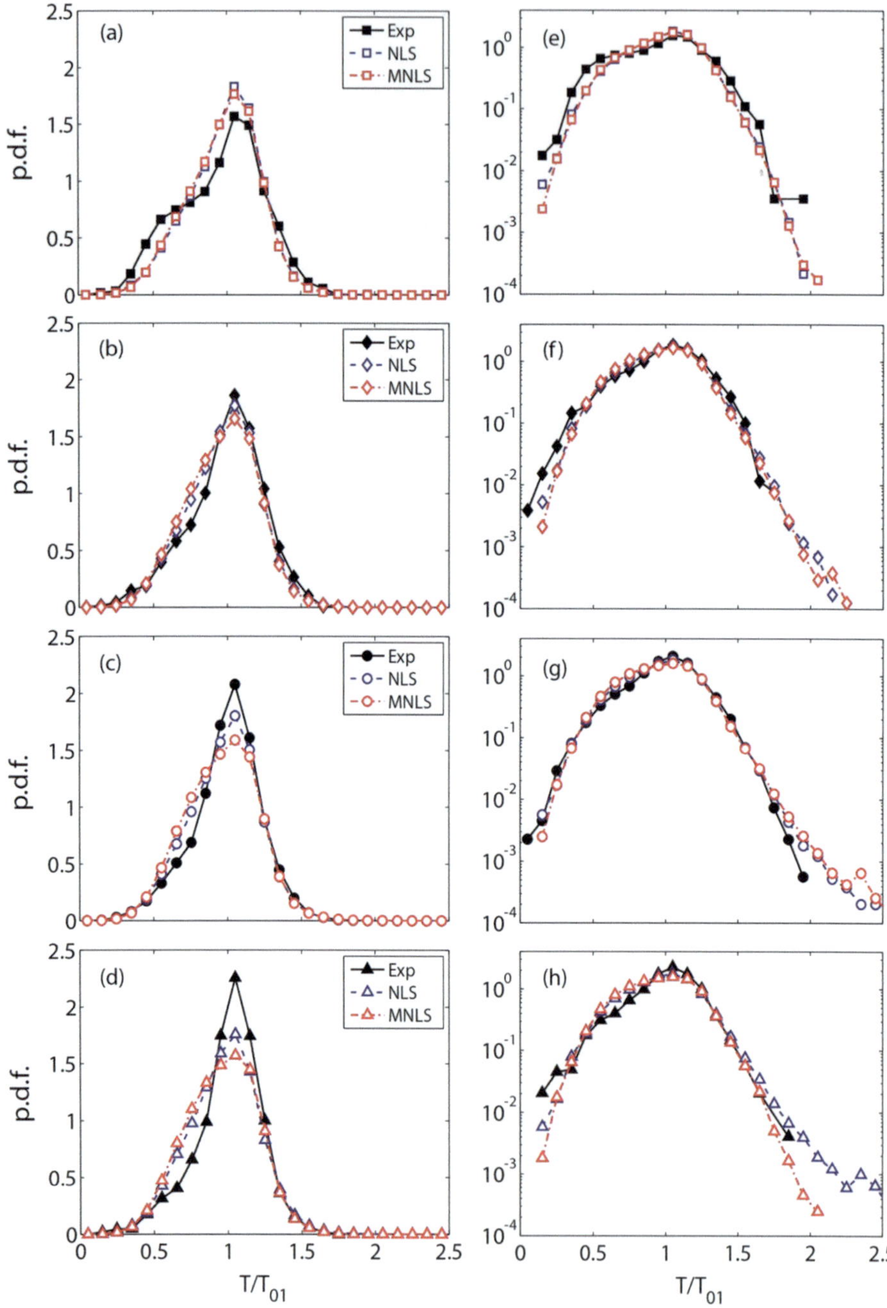

Fig. 4.15 **a–d** Probability density function of wave period in the same location of Gauge 8 for different sea states. **e–h** are the same distributions but on a logarithmic scale

the right column of Fig. 4.15. The slight variation of the tail among different sea states indicates that the modulational instability enlarges the magnitude of the wave crest without apparently increasing the associated wave length. Finally, what needs to be pointed out is that the spatial variation of wave period in numerical simulations can be neglected as well.

In closing, it must be mentioned that these waves, which have been generated in a wave basin, will have slightly different characteristics from those of the full-scale waves in the ocean and that although similar kinds of studies with ocean waves are possible (Slunyaev et al. 2005, 2013), the availability of such measured data and the control about its characteristics are very limited.

It should also be mentioned that real ocean waves are not long-crested, and directional spreading can play an important role in the determination of the statistical properties of the surface elevation (Onorato et al. 2009; Mori et al. 2011).

4.4 Conclusion

Firstly, the evolution of a propagating wave group with an initial Gaussian-shaped envelope has been exploited. Particularly, the effect of each four-order term has been considered separately. It is revealed that the extreme wave heights are primarily attributed to the third-order nonlinear effect, which confirms again that the NLS equation can catch the essential characteristics of the extreme waves. The emerged asymmetry of the envelope is mainly due to the result of the first fourth-order nonlinear term IV in the Dysthe model (See Eq. (4.1)), and the induced mean current represented by term VI has a negative effect on the energy focusing process, suppressing the formation of extreme waves. No significant contribution can be found for the rest of the fourth-order nonlinear terms.

Comparing the numerical simulations performed by NLS and Dysthe equations with laboratory experiments in Marintek, it is found that the vertical asymmetry of the surface elevation provoked by the second-order nonlinear effect does not show a significant difference between the two numerical simulations except for a little overestimation on experimental observation. However, the nonlinear modulation, i.e., a quasi-resonant four-wave interaction process that takes place near the spectrum's peak, does lead to a significant divergence in the coefficient of kurtosis.

Since there is no energy dissipation in numerical simulations and a stronger energy focusing in the NLS model, the coefficient of kurtosis obtained from the NLS equation not only achieves the fully developed condition earlier but also presents a higher value than those in Dysthe simulation and laboratory experiment. However, nearly identical spatial evolutionary tendencies can be observed between rogue wave density and the coefficient of kurtosis, confirming again that the kurtosis coefficient can describe the appearance of abnormal waves.

In terms of the exceedance distribution of wave height, except for the intermediate stage of the wave evolutionary process, both NLS and Dysthe models can catch the formation of extreme waves very well if the spatial evolutionary distance is not very long. More precisely, in the low sea state, both numerical simulations and laboratory experiments tend to the linear Rayleigh distribution due to the dominant dispersive effect; in the moderate and severe sea states, the evolutionary process changes from linear to nonlinear and shows a good agreement with experiments at the initial and fully developed stages; the discrepancy cannot be avoided in the intermediate stage as a result of the stronger nonlinearity in simulation and the increased phenomenon of wave breaking in reality.

With the increased nonlinearity, the wave surface elevations begin to deviate from the Gaussian process, becoming narrower and positively skewed in the domain close to the mode of distribution. Comparisons with various theoretical models indicate that the deviation in this range, except for the peak value, is mainly attributed to the bound wave effects, while for the larger amplitude of surface elevation, especially in the positive part, the difference is principally caused by the modulational instability. Moreover, the NLS-type simulations are perfectly consistent with the experimental results of up to 4 standard deviations in all sea states.

The second-order interaction participates in the deviation from Gaussian statistics. If the nonlinearity is weak, the wave crest distribution can be fitted well by both second-order and third-order nonlinear models due to the limited nonlinear effects. Once the nonlinearity increases, e.g. in the moderate sea state, the nonlinear instability will immediately dominate the evolutionary process and generate a larger amplitude of crest, which can only be described by the third-order GC model. Similarly, as observed in wave height distribution, the extreme crest is determined by the number of waves in the series. Furthermore, a linear regression model exists between the scaled maximum wave crest and the coefficient of kurtosis, indicating that the abnormal wave appears typically with a very large wave crest.

A similar spatial variation for wave trough is also detected in the same sea state, but it is not as sensitive as wave crest to nonlinear instability. The quasi-resonant four-wave interaction largely dominates the evolution of wave profile, generating a sharper and higher crest easily and creating a deeper trough when the nonlinearity is fully developed, which can be indicated by an approximately stable kurtosis coefficient value.

Notably, the two numerical simulations are still in agreement with each other and fit the observed statistics of wave trough reasonably well in most cases. The major discrepancy between numerical simulations of NLS and Dysthe models exhibited in the wave height distribution is primarily attributed to the difference in wave crest distribution.

The spatial variation of the wave period is not very obvious except for becoming a little narrower downstream along the wave tank. Moreover, the period distributions in a wave train are much narrower than that of wave crests or troughs, and the spread lies mainly in the range from 0.5 to 1.5 times the mean wave period. With the increased nonlinearity in these sea states, the observed wave period distribution becomes steeper and narrower, while the numerical simulations cannot clearly identify this tendency. However, the main characteristics are still reflected correctly by NLS-type equations. The little variation on the right tail of wave period distribution reveals that the nonlinear instability does not enlarge wave length noticeably.

Modelling Extreme Waves with 3D Dysthe Equation

5

5.1 Basic Theory

5.1.1 Temporal and Spatial Versions of 3D MNLS Equation

The evolution of a wave group can be represented by the variation of the complex amplitude of either the surface elevation or the velocity potential at the free surface in the time or space domain. In this chapter, the equations expressed in the dimensionless complex amplitude of the surface elevation are employed to simulate the evolution of the wave field. The exact expressions for the temporal and spatial 3D MNLS equations are given by (Socquet-Juglard et al. 2005),

$$
\begin{aligned}
&B_t + \frac{1}{2}B_x + \frac{\mathrm{i}}{8}B_{xx} - \frac{\mathrm{i}}{4}B_{yy} + \frac{\mathrm{i}}{2}|B|^2B - \frac{1}{16}B_{xxx} + \frac{3}{8}B_{xyy} \\
&\quad + \frac{3}{2}|B|^2B_x + \frac{1}{4}B^2B_x^* + \mathrm{i}B\overline{\phi}_x = 0, \quad z = 0,
\end{aligned}
\tag{5.1}
$$

$$
\begin{aligned}
&B_x + 2B_t + \mathrm{i}B_{tt} - \frac{\mathrm{i}}{2}B_{yy} + \mathrm{i}|B|^2B - B_{tyy} - 8|B|^2B_t \\
&\quad - 2B^2B_t^* - 4\mathrm{i}B\overline{\phi}_t = 0, \quad z = 0,
\end{aligned}
\tag{5.2}
$$

where the induced mean flow potential $\overline{\phi}$ is governed by

$$\nabla^2\overline{\phi} = 0, \quad -h < z < 0, \tag{5.3}$$

$$\overline{\phi}_z = \frac{1}{2}\left(|B|^2\right)_x = -\left(|B|^2\right)_t, \quad z = 0, \tag{5.4}$$

$$\overline{\phi}_z = 0, \quad z = -h. \tag{5.5}$$

H. Zhang and C. Guedes Soares, *Numerical Modelling of Extreme Waves*, Synthesis Lectures on Ocean Systems Engineering, https://doi.org/10.1007/978-3-031-77084-5_5

Here the subscripts denote partial derivatives. Time and length are scaled by peak frequency ω_p and corresponding peak wave number k_p, respectively. If the free wave complex amplitude B is solved, the surface displacement can be reconstructed by the following formula:

$$\eta = \overline{\eta} + \frac{1}{2}\left(B\mathrm{e}^{\mathrm{i}(x-t)} + B_2\mathrm{e}^{2\mathrm{i}(x-t)} + B_3\mathrm{e}^{3\mathrm{i}(x-t)} + \cdots + \mathrm{c.c.}\right), \tag{5.6}$$

where c.c. denotes complex conjugate, and the bound contributions are given by

$$\overline{\eta} = \frac{1}{2}\overline{\Phi}_x = -\overline{\Phi}_t, \tag{5.7}$$

$$B_2 = \frac{1}{2}B^2 - \frac{\mathrm{i}}{2}BB_x = \frac{1}{2}B^2 + \mathrm{i}BB_t, \tag{5.8}$$

$$B_3 = \frac{3}{8}B^3. \tag{5.9}$$

Both B_2 and B_3 are phase-locked to B and the mean surface elevation $\overline{\eta}$ is associated with the radiation stress.

5.1.2 Initial Random Multidirectional Wave Field

The JONSWAP spectrum specifies the initial wave field with a directional spreading function.

$$S(k,\theta) = \frac{\alpha}{2k^3}\exp\left[-\frac{5}{4}\left(\frac{k_p}{k}\right)^2\right]\gamma^{\exp\left[-\left(\sqrt{k/k_p}-1\right)^2 \Big/ \left(2\sigma_0^2\right)\right]}D(\theta), \tag{5.10}$$

where k_p is the peak wave number and the parameter σ_0 has the standard values: 0.07 for $k \le k_p$ and 0.09 for $k > k_p$. The Philips parameter α is related to the desired significant wave height and the peak enhancement factor γ specifies the spectral bandwidth.

The directional spreading is taken to be of the form:

$$D(\theta) = \begin{cases} \cos^N(\theta) \big/ \Omega(N), & |\theta| \le \pi/2, \\ 0, & \text{otherwise}, \end{cases} \tag{5.11}$$

where $\Omega(N)$ is the normalizing factor of the directional distribution function, and N is a measure of the directional spreading, different values of which can be used, ranging from fairly long-crested (large N) to short-crested (small N) waves. To be consistent with the experimental results, the following values have been selected in numerical simulations: $N = 840$, 200, 90, 50 and 24, as demonstrated in Fig. 2.29.

Now in the temporal evolution, the corresponding directional wave number spectrum can be derived by $S(\boldsymbol{k}) = S(k, \theta) / k$, and the computations are initiated by specifying the spatial Fourier transform of B at $t = 0$,

$$\widehat{B}(\boldsymbol{K}_{mn}, 0) = \sqrt{-2S(\boldsymbol{K}_{mn})\Delta K_x \Delta K_y \log(\hat{\theta}_{mn})} \exp(\mathrm{i}2\pi\theta_{mn}), \tag{5.12}$$

where $\hat{\theta}_{mn}$ and θ_{mn} are independently taken to be uniformly distributed on [0, 1), and $\boldsymbol{K}_{mn} = (m\Delta K_x, n\Delta K_y)$. The physical amplitude $B(\boldsymbol{x}_{jk}, t)$ is obtained through the following inverse discrete Fourier transform:

$$B(\boldsymbol{x}_{jk}, t) = \sum_{m=-N_x/2}^{N_x/2} \sum_{n=-N_y/2}^{N_y/2} \widehat{B}(\boldsymbol{K}_{mn}, t) \exp(\mathrm{i}\boldsymbol{K}_{mn} \cdot \boldsymbol{x}_{jk}). \tag{5.13}$$

With this initialization method, each Fourier coefficient can be an independent complex Gaussian variable with a variance proportional to the corresponding value of the spectrum. The preparation of the initial condition in the spatial evolution is almost the same. The initial wave spectrum $S(\omega, k_y)$ can be obtained directly from $S(\boldsymbol{k})$ by using the Jacobian transform, and one more condition has to be added in this case, that is, $\omega^4 > g^2 k_y^2$.

5.2 Numerical Schemes

The laboratory experiments performed at the Marintek wave basin will be simulated, and detailed descriptions of the facilities and experiments can be found in the earlier works (Cherneva et al. 2009; Onorato et al. 2009; Toffoli et al. 2010a). Here, only a brief introduction is given to help illustrate how the numerical tests are designed and compared with the corresponding physical experiments of multidirectional waves.

Waves are generated in a large rectangular wave basin with dimensions of 80 m × 50 m as presented in Fig. 5.1. This basin is equipped with two sets of wavemakers. One is a double flap fixed along the 50 m side, which is a hydraulically operated unit for generating long-crested, regular and irregular waves. The other one used for the present study is installed along the 80 m side, consisting of 144 individually computer-controlled flaps and generating short-crested seas within a wide range of directional wave energy distributions. The water depth can also be changed by moving the bottom of the basin up or down, and it was fixed at 3 m for the present experiments, i.e. $k_p h = 12.07 > 3.14$, satisfying the deepwater condition for both cases in Table 2.11. The absorbing sloping beach is also equipped on the opposite side to reduce the wave reflections.

To trace the evolution of the wave field, wave measurements have been mainly concentrated along the central line of the basin (see Fig. 5.1). The temporal variation of the surface displacement is recorded with a sampling frequency of 80 Hz by the wave probes uniformly deployed at 5 m intervals. Considering that the experiments were carried out in

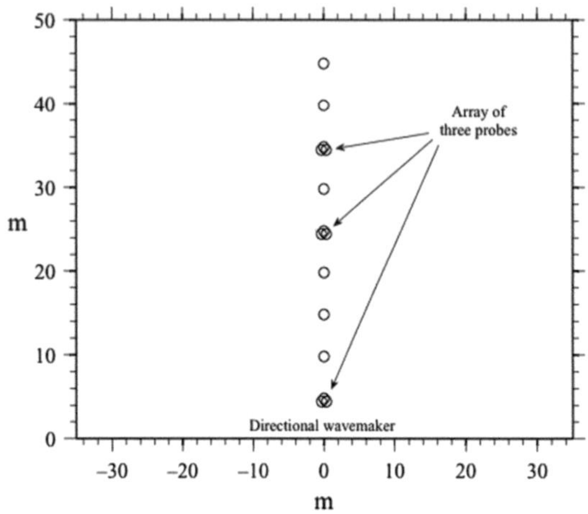

Fig. 5.1 Layout of wave gauges in Marintek

a finite-size basin, the results may be affected by the discreteness of the facility. However, as illustrated in the work of Zhang et al. (2013), the aspect ratio of the facility cannot significantly affect the wave statistics.

As to the initial conditions at the wavemaker, the complex Fourier amplitudes are generated, each with its modulus randomly chosen from a Rayleigh distribution around the target spectrum, and random phases assumed to be uniformly distributed between 0 and 2π.

In the present study, two different types of experiments are carried out, characterized by two distinct groups of Philips parameter α and peak enhancement factor γ as listed in Table 2.11, where *BFI* is also calculated considering that this parameter contains information of both wave steepness and spectral bandwidth. To have enough samples to produce a significant statistical analysis, four realizations of the random sea surface for a given spectrum were performed using different sets of random amplitudes and phases. For each test, 20 min of wave records are collected, including the initial ramp-up time that has to be cut off in the analysis.

The split-step Fourier method described by Lo and Mei is used to solve the MNLS equations numerically in a three-dimensional rectangular domain with an imposed periodic boundary condition in both horizontal directions. For example, in the temporal evolution, the envelope function $B(\boldsymbol{x}, t)$ is solved in the modulational wavenumber space $\boldsymbol{K} = (K_x, K_y) = (\boldsymbol{k} - \boldsymbol{k}_p) / k_p$ and only the modes within $|K_x| \leq 1$ and $|K_y| \leq 1$ are used. The computation domain is $256L \times 256L$, over which a uniform grid size $N_x = N_y = 512$ is applied. To resolve the rapidly oscillating surface elevation $\eta(\boldsymbol{x}, t)$, a resampling is performed. For the resolution explained above, the minimum number of points that must be used to describe the full surface elevation, including the third-order bound wave effect, is 3072×1536, corresponding to about 12 points per dominant wavelength L downstream in the wave tank.

A similar setup of these parameters is also applied to the spatial evolution. A detailed description of the numerical method can be found in the work of Socquet-Juglard et al. (2005). Considering that the free wave envelop B is slowly varying under the narrowband assumption, the integration step used in the numerical simulation can be chosen relatively large (Toffoli et al. 2010a) such as $\Delta t = \Delta x = 2\pi / 10$. It corresponds to 10 integration steps per peak period T in the temporal evolution and 10 integration steps per peak wavelength L in the spatial evolution. These parameters can provide satisfactory accuracy, and the energy in all simulations is verified to be conserved to within 0.4% of the initial value.

In numerical simulation, only the model based on the spatial version of the MNLS equation is consistent with the laboratory tests, i.e., imposing the boundary condition at $x = 0$ and obtaining the spatial evolution of the temporal wave series along the basin.

However, the temporal version of the MNLS equation provides the temporal evolution of an initial wave field imposed at $t = 0$, and thus, some problems will arise when comparing the numerical results with the laboratory experiments measured by wave gauges.

In the following analysis, the comparison is based on the leading-order approximation that all properties are functions of $x - c_g t$, where c_g is the wave group velocity. Hence, the results at location $x = L$ in the spatial evolution are assumed to correspond to those at time $t = 2T$ in the temporal evolution. Note that L and T represent the dominant wavelength and peak period, respectively. This assumption has been simply checked in the work of Toffoli et al. (2010a) and will be further considered in this chapter.

Simulations with the spatial MNLS equation are run up to $x = 30L$, with 9 outputs corresponding to the locations of wave gauges in the basin. Correspondingly, using the above-mentioned approximation to relate the spatial and temporal evolution, the total duration of the temporal MNLS simulations has to be set equal to $t = 60T$.

The output surface elevation is used to calculate the statistical properties of the wave field. To achieve statistically significant results, 10 realizations have been performed with the same input spectrum and different random amplitudes and phases (Xiao et al. 2013). Besides, a large number of random realizations are expected to minimize the influence related to the discrepancy in initial conditions.

5.3 Difference Between Temporal and Spatial Evolutions

Shemer and Dorfman (2008) conducted an experimental and numerical study of nonlinear wave groups' spatial and temporal evolution in two dimensions. To provide an initial intuition on the properties of these two kinds of evolution, the associated numerical experiments have been further developed herein and performed to a much longer distance.

To the leading order, the initial surface elevation at the wavemaker has the same form as Eq. (3.12), of which the forcing amplitude and carrier wave period are selected from

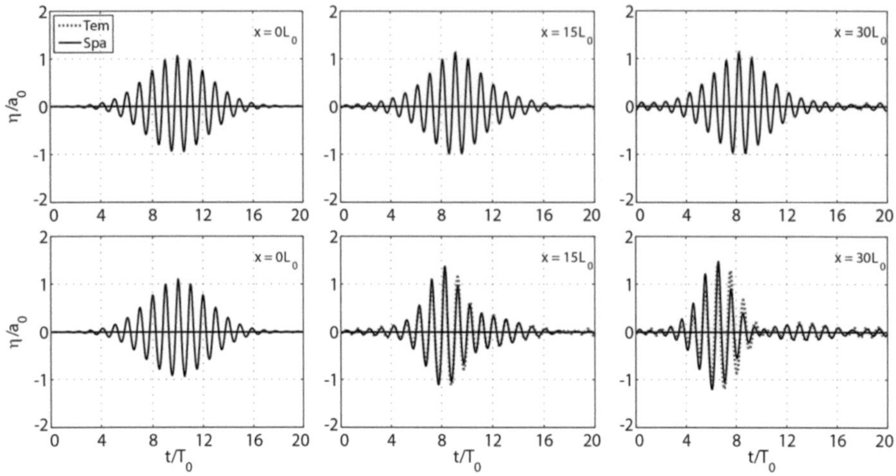

Fig. 5.2 Temporal variations of surface elevation obtained at different locations simulated by the spatial evolution model (black solid line), comparing with those obtained indirectly from the temporal evolution model (red dotted line). $a_0k_0 = 0.13$, first panel, and $a_0k_0 = 0.16$, second panel

Table 2.11 in purpose such that $a_0 = H_s/2$ and $T_0 = T$. Note that in Fig. 5.2, L_0 is the carrier wavelength calculated from the linear dispersion relation. The value of $m = 3.5$ is chosen, being sufficient to satisfy the spectrum constrain for applying the MNLS equation (Shemer et al. 2002).

The initial wave envelope $B(t, x = 0)$ is reconstructed on the basis of time series represented by Eq. (3.12). Then the wave envelope is simulated upstream by the spatial version of the MNLS equation (e.g., Eq. (5.2)) up to a specified location. Hence the wave fields $B(x, t)$, $\eta(x, t)$ are obtained in some domain (x, t) and the initial conditions $B(x, t = 0)$, $\eta(x, t = 0)$ can be formulated and used to work as the initial condition for the temporal version of MNLS equation (e.g., Eq. (5.1)). Detailed explanation of this procedure can be found in the work of Slunyaev et al. (2005).

In the first row of Fig. 5.2, the temporal and spatial evolutions of the surface elevation in a wave group with an initial Gaussian-shaped envelope do not present any difference for the reason that the nonlinearity is weak, indicated directly by the slow variation of the modulational process.

However, with a much larger initial wave steepness in the second row, the discrepancy between temporal and spatial evolutions becomes increasingly evident as the waves propagate downstream in the wave basin. The temporal evolution model seems to present a larger value than the spatial evolution model. Moreover, it is evident that both groups extend in the horizontal direction, and the initially symmetric Gaussian wave packet gradually exhibits stronger front-tail asymmetry, with increasingly steep front and elongated tail. The crest-trough asymmetry is also notable and mainly attributed to the bound wave effects.

The difference indicated above is only based on a two-dimensional wave group. More general three-dimensional waves with an initial JONSWAP spectrum will be analyzed and compared with experimental results in the following parts.

5.4 Numerical Simulation of Multidirectional Waves

5.4.1 Coefficients of Skewness and Kurtosis

From Figs. 5.3, 5.4, 5.5 and 5.6, the coefficients of skewness and kurtosis are plotted as a function of the dimensionless distance to the wavemaker, i.e., dividing x by the wavelength L corresponding to the initial peak period T. With regard to the error bars, the upper and lower error ranges of each point are equal to the mean value plus and minus its standard deviation, respectively (Zhang et al. 2016b).

It is well known that the nonlinear interaction between wave components can induce a vertically asymmetric profile with a higher, more peaked crest and shallower, more rounded trough. This deviation is mainly dominated by the bound wave effects, even though the dynamics of free wave weakly contribute (Onorato et al. 2005; Mori et al. 2007). The coefficient of skewness shows a slightly decreasing tendency along the wave basin, partly due to the spectral downshift that reduces the steepness and, consequently,

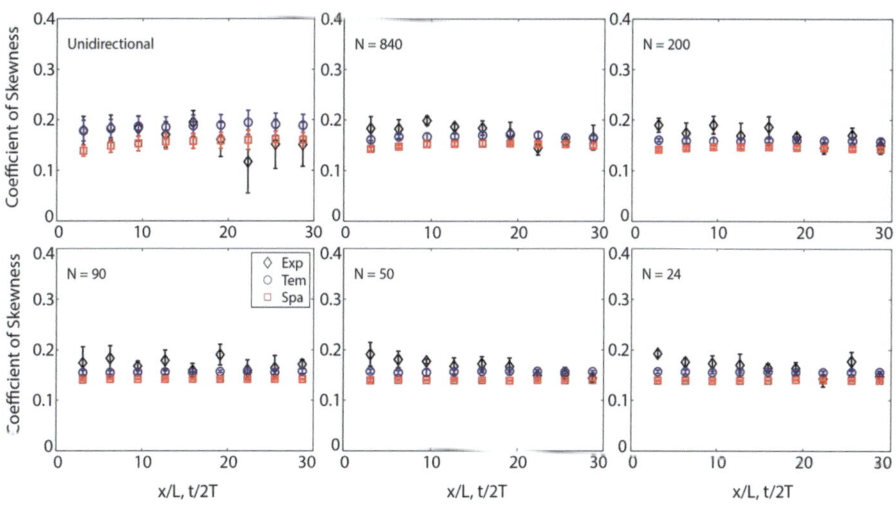

Fig. 5.3 Evolution of coefficient of skewness for $BFI = 0.70$

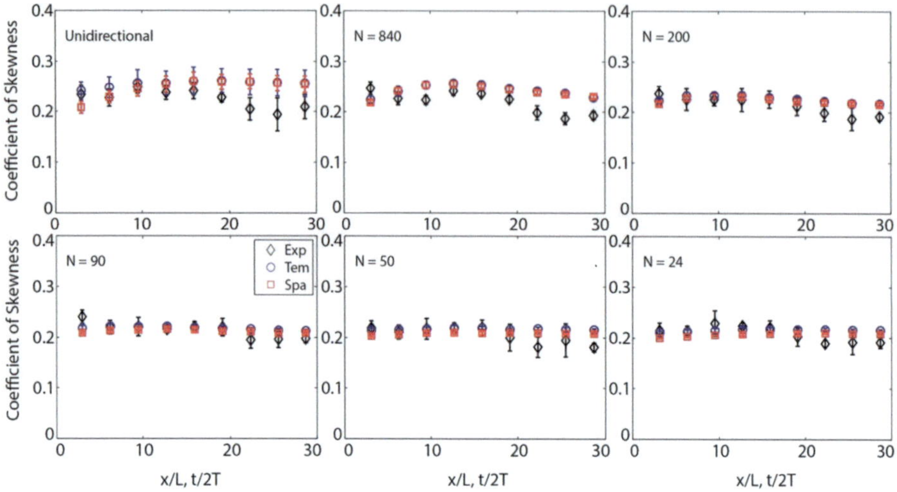

Fig. 5.4 Evolution of coefficient of skewness for $BFI = 1.10$

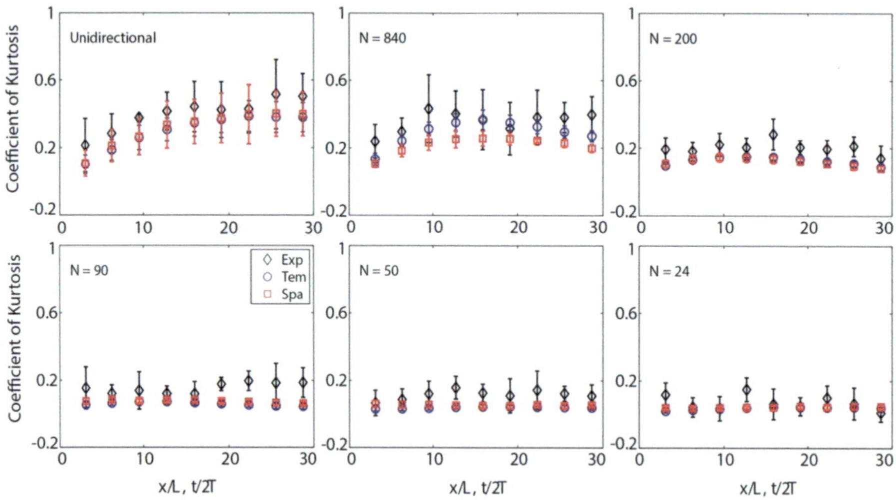

Fig. 5.5 Evolution of coefficient of kurtosis for $BFI = 0.70$

the contribution of bound waves (Toffoli et al. 2010a). The numerical simulations partly capture this trend, such as in the case of $N = 840$ in Fig. 5.4.

Although a little discrepancy can be observed between numerical simulations and laboratory experiments, both models have caught the fundamental characteristics of skewness. Moreover, the asymmetry of the wave profile in temporal evolution is a little larger than that presented in spatial evolution in the cases with initial $BFI = 0.70$. In summary, it

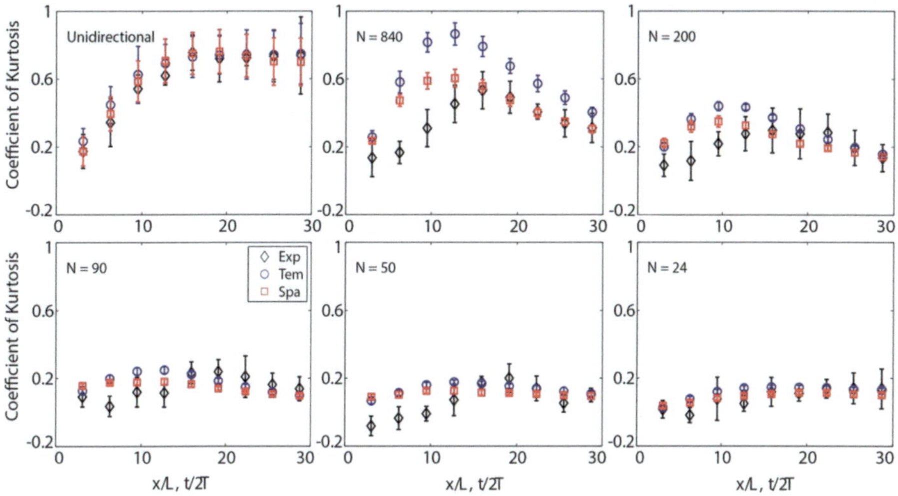

Fig. 5.6 Evolution of coefficient of kurtosis for $BFI = 1.10$

can be argued that directional spreading does not lead to any distinct difference in the statistical property associated with vertical asymmetry.

Compared with the skewness, the coefficient of kurtosis can deviate substantially from normality due to the nonlinear instability, which gives rise to large amplitude waves (Onorato et al. 2006). Such a phenomenon is very pronounced in long-crested wave fields, as presented in Figs. 5.5 and 5.6, where the statistical uncertainty indicated by the length of error bar is also very large.

In agreement with the previous laboratory experiments (Onorato et al. 2005), it reaches its maximum after a reasonably short evolution equivalent to about 15–20 wavelengths and then decreases towards the end of the basin with various degrees, depending on the initial *BFI* and directional spreading *N*.

For a broad directional distribution, e.g., $N \leq 90$, the nonlinear instability no longer significantly contributes to the statistical properties of waves. Thus, the kurtosis only weakly deviates from Gaussian statistics, which both versions of numerical models can capture, although a little discrepancy still could be observed.

Concerning the long-crested waves, e.g., $N > 200$, if the initial *BFI* is small, such as Fig. 5.5 in this section, the temporal evolution is almost in agreement with the spatial one, and both numerical models can estimate the dynamics of mechanically generated waves reasonably well. However, as the initial *BFI* increases (see Fig. 5.6), the temporal version of the MNLS equation overestimates the coefficient of kurtosis (e.g., $N = 840$) while the spatial model is still in agreement with the experimental result, except for a small overshoot at the initial stage, which disappears surprisingly in the unidirectional case.

5.4.2 Surface Elevation

In this section, the study is focused on the surface elevation, which has been scaled by the standard deviation σ of the concurrent wave series. Both numerical and experimental probability density functions are compared with the Gaussian distribution and the Tayfun second-order model, approximated more concisely by Socquet-Juglard et al. (2005).

In the case of unidirectional waves, i.e., Fig. 5.7, the experimental probability density function only fits the Tayfun second-order distribution reasonably well at the first probe ($x/L = 3.1$) where the statistical properties of the wave field are only dominated by the bound waves. As the waves propagate along the basin, a substantial deviation is observed on both tails of the distribution, which can be mainly attributed to the nonlinear dynamics of free wave modes.

Amazingly, the lower tail of the probability density function can be described well by the normal distribution, which agrees with the results of the previous analysis of the Marintek dataset in Sect. 4.3.2. Both versions of the MNLS equation can work very well and are consistent with the experimental results in all locations regardless of the initial *BFI*.

Concerning the short-crested waves, e.g., Figs. 5.8 and 5.9, it is pronounced that the departure of the tail of the probability density function from the second-order theoretical model is largely reduced.

In other words, the coexistence of a large number of wave components with different propagating directions leads to a huge reduction on the effects of the nonlinear instability,

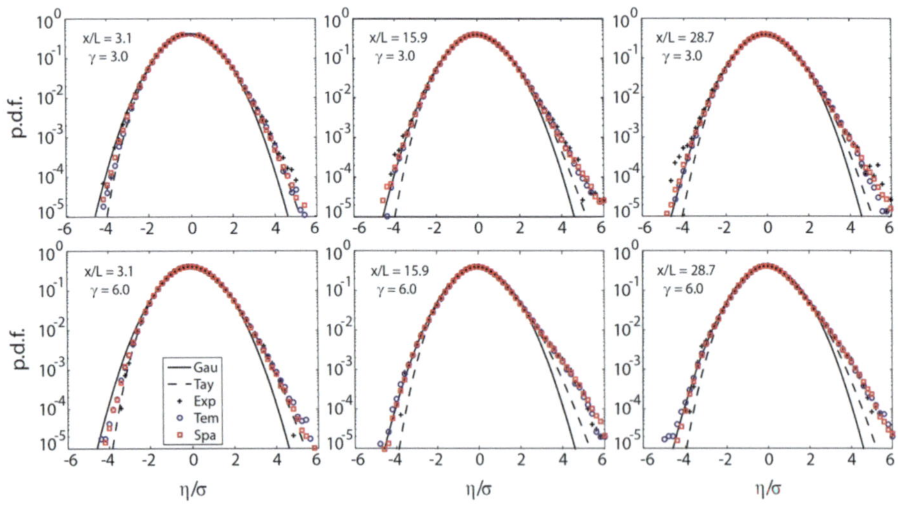

Fig. 5.7 Probability density function of surface elevation for unidirectional waves

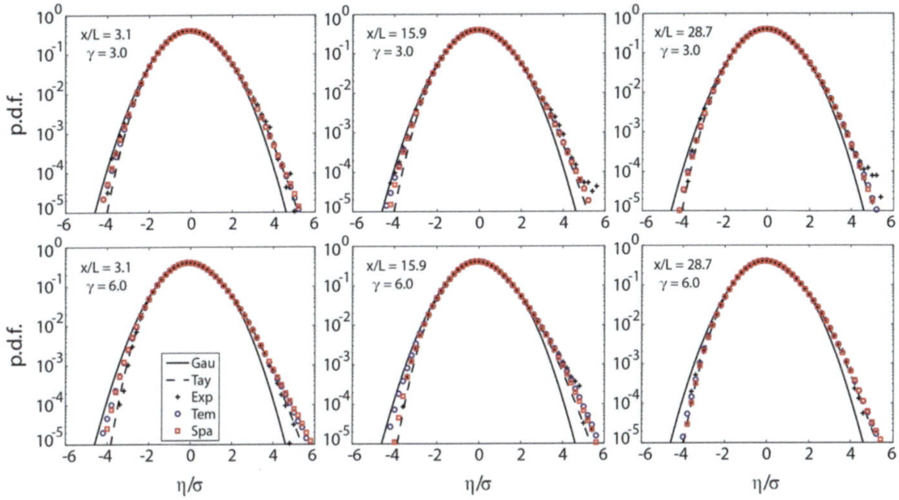

Fig. 5.8 Probability density function of surface elevation with initial directional spreading $N = 200$

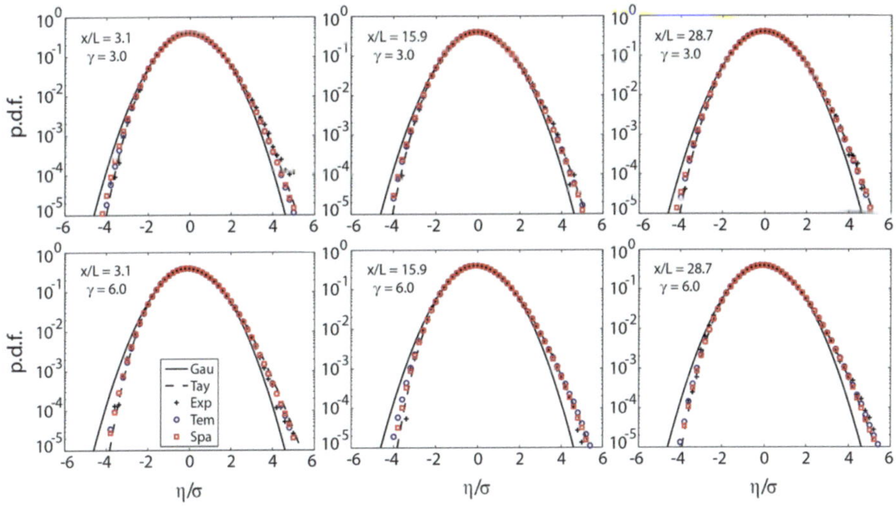

Fig. 5.9 Probability density function of surface elevation with initial directional spreading $N = 24$

even in the case of a larger initial *BFI*. The discrepancy between spatial and temporal simulations can be neglected, and both governing equations can capture the aforementioned deviations very well, even though they are not fully nonlinear wave models.

Another very important nonlinear property of surface elevations, i.e. being positively skewed, can be observed in Fig. 5.10 and fitted well by the second-order model. The

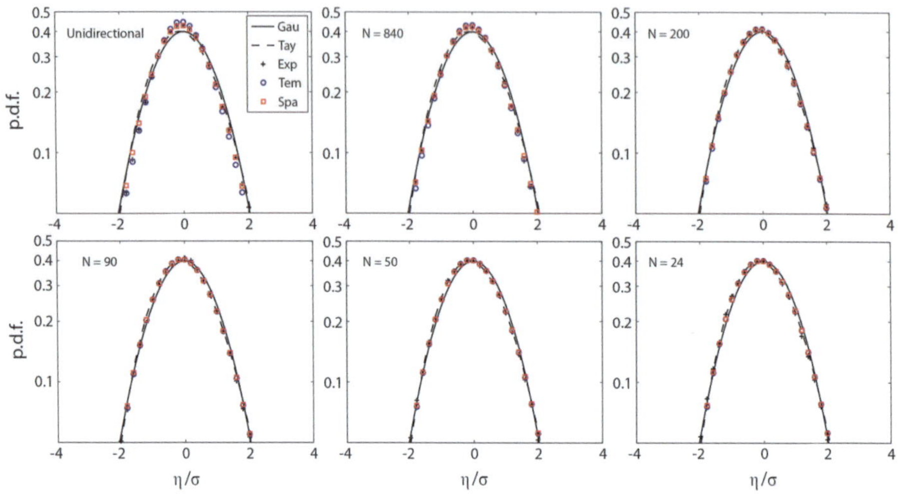

Fig. 5.10 Probability density function of surface elevation located at $x/L = 15.9$ with the same $\gamma = 6$ and different initial directional spreading N

bound wave effects induce this phenomenon. Furthermore, the peak value of the probability density function is larger than the second-order prediction only in the case of long-crested waves. Hence, it can be inferred that this kind of deviation should be mainly attributed to the nonlinear instability induced by the quasi-resonant or resonant four-wave interaction, which is represented very well by both versions of MNLS equations.

As shown in Figs. 5.11 and 5.12, the averaged maximum surface elevations in temporal evolution are consistent with those obtained in spatial evolution in most cases. Comparing with the laboratory observations, the agreement is still reasonably good for unidirectional waves, as indicated in Fig. 5.11, while the difference is evident in directional sea states, as presented in Fig. 5.12, especially in the case with larger initial *BFI* ($\gamma = 6$). It is reasonable, considering that the surface elevation is only measured at one point in the wave basin.

Since these probes are arranged in the central line, the large discrepancy reveals that the maximum surface elevation appears outside the middle line of the wave basin, which may be explained by the increased mean directional spread in the evolutionary process (Xiao et al. 2013).

However, the averaged maximum surface elevations in three-dimensional waves can still be evaluated within the framework of Piterbarg's theory (Socquet-Juglard et al. 2005). Except for the case of $N = 840$, the theoretical estimations (solid and dash lines) have given a good prediction of the numerical results.

It has been reported that the maximum kurtosis in the directional wave field will decrease as the initial mean directional spreading increases (Toffoli et al. 2010a), and a similar tendency is discovered for the extreme surface elevation as illustrated in Fig. 5.13.

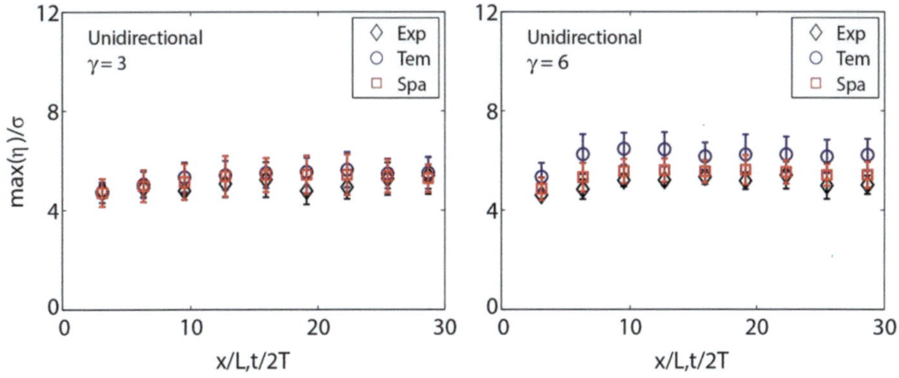

Fig. 5.11 Evolution of the maximum surface elevation in unidirectional sea state

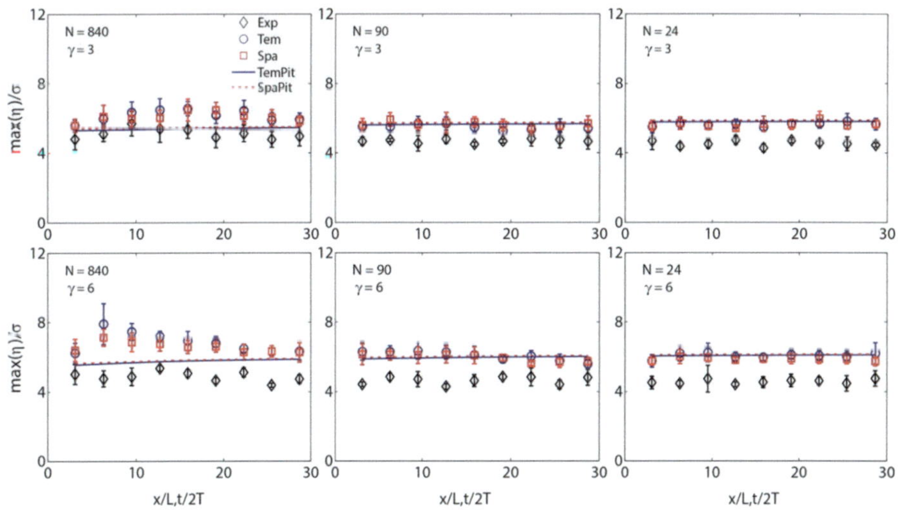

Fig. 5.12 Evolution of the maximum surface elevation with different initial directional spreading N, compared with the expected value given by Piterbarg–Tayfun distribution (solid and dash lines)

Thus, to a certain degree, it confirms that the coefficient of kurtosis can work as an indication of the formation of the extreme crest in three-dimensional water waves (Guedes Soares et al. 2003, 2004), as well as in two-dimensional cases (Zhang et al. 2015).

To compare with the ideal long-crested waves, the unidirectional case has been denoted as $N = 2000$ in Fig. 5.13. What is worth mentioning is that the maximal extreme surface elevation appears in the case of $N = 840$ rather than in the unidirectional waves in both numerical simulations and the first experimental result.

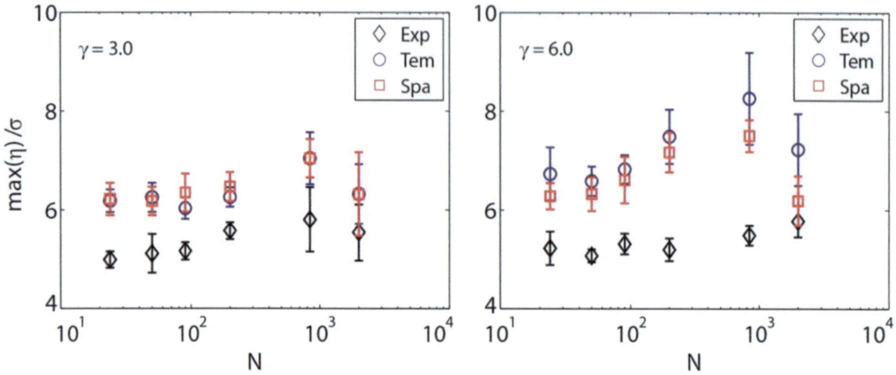

Fig. 5.13 Extreme surface elevation as a function of initial directional spreading N. Unidirectional wave is denoted as 2000 for comparison

As expected, the discrepancy between numerical simulations and laboratory experiments is pronounced in three-dimensional cases due to the different sizes of the observed domain and individual waves, and some of the theoretical methods for predicting the extreme surface elevation in a specified sea state can be found in the previous works where the correlation between space and time can be taken into account.

5.4.3 Wavenumber Spectrum

The statistical property of surface elevation has been compared and discussed in Sect. 5.4.2. However, due to the limitation of wave measurement in 3D wave tank, the faithful reconstruction of the wave number spectrum is impossible. Therefore, in this section, only the temporal evolution of the wave number spectrum in numerical simulation is briefly discussed for case B with $N = 840$.

One example of the surface elevations of three-dimensional long-crested waves is plotted in Fig. 5.14 where the "great water wall" can be observed, suggested by the large height and transverse width of the peak crest, which are in agreement with the observation in the natural ocean environment reported by crew members.

To demonstrate the evolutionary tendency of the wave number spectrum, Case B with $N = 840$ has been taken as an example, of which the temporal evolution of the directional as well as the directionally integrated wave number spectra are shown in Figs. 5.15 and 5.16, respectively. These figures reveal a significant spectrum evolution that occurs over time, mainly indicating its width, which broadens gradually. Moreover, a small downshift on the peak frequency and a tendency towards a $k^{2.5}$ power law in the high wave number spectral tail can be clearly observed. In addition, even though it is not presented here, it has to be mentioned that for case A (smaller *BFI*) with the same initial directional spreading N,

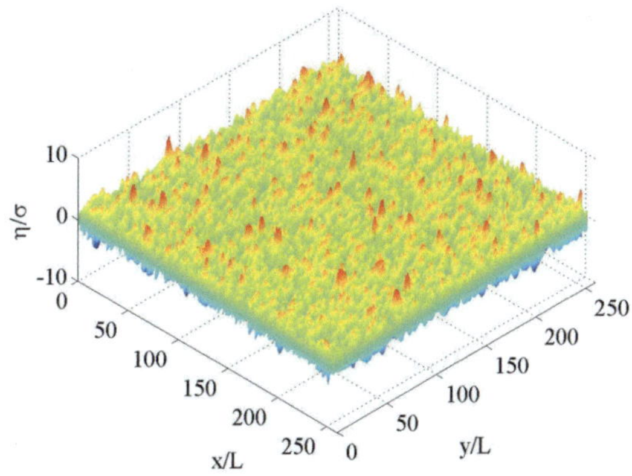

Fig. 5.14 Surface elevation of three-dimensional long-crested waves in temporal evolution

the degree of broadening will be much lower, meaning that the strong nonlinearity in the wave filed typically can result in a broader wave spectrum.

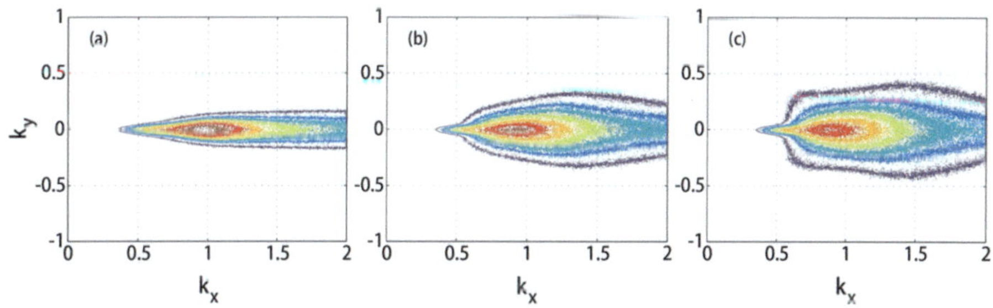

Fig. 5.15 Temporal evolution of wave number spectrum for case B with $N = 840$. **a–c** are the evolutions at the instant of 6.2 T, 31.8 T and 57.4 T, respectively

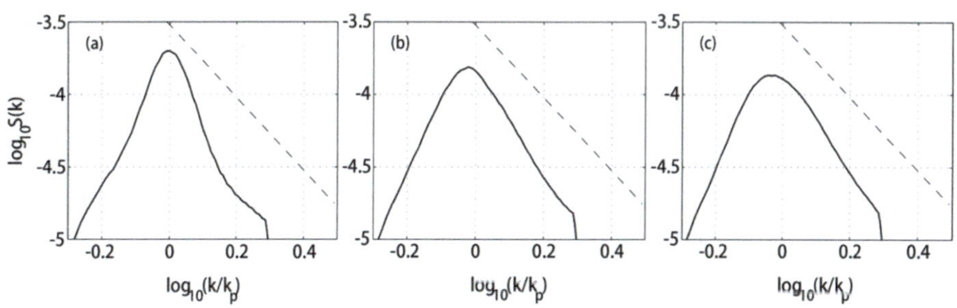

Fig. 5.16 The same wave number spectrum as Fig. 5.15 but for the modulus. The dashed line is the power law $k^{2.5}$

Last, it has to be mentioned that similar conclusions have been reported for the initial Gaussian-shaped spectrum (Dysthe et al. 2003) and the same JONSWAP wave spectrum but with different directional spreading functions (Socquet-Juglard et al. 2005). All the numerical results herein are qualitatively consistent with the laboratory observations in the work of Onorato et al. (2009), where arrays of wave gauges were deployed in the tank, and the wavelet directional method is used to estimate the directional distribution.

5.5 Conclusion

For a 2D Gaussian wave packet, both versions of numerical models can catch the critical feature of the gradual change of an initially symmetric shape into a strongly asymmetric one. However, in the presence of strong nonlinearity, the temporal evolution is not always in agreement with the spatial one and can present a large discrepancy in the peak stage of the evolutionary process, which may be partially attributed to the tiny difference in the dispersion relation.

A slightly decreasing tendency can be observed for the coefficient of skewness in both 2D and 3D laboratory experiments, partially due to the spectral downshift that reduces the steepness and, consequently, the contribution of bound waves. Although a little discrepancy is seen between numerical simulations and experimental results, both wave models can catch the fundamental characteristics of asymmetry of wave profiles.

Regarding the coefficient of kurtosis, some discrepancies are detected in the case with narrow directional spreading $N = 840$ and large initial *BFI*, performing an overestimation on the kurtosis at short distances to the wavemaker, particularly for the temporal evolution mode. However, this discrepancy is insignificant in unidirectional waves, as presented in the numerical simulations governed by 2D spatial and temporal MNLS equations. Both numerical models estimate this statistical variable's evolution reasonably well for the sea states with broader directional spreading, independent of initial BFI.

Concerning the probability density function of surface elevation, the two versions of numerical models can capture the deviation on both tails that the second-order theory in a long-crested sea state cannot describe. Furthermore, in the case of short-crested waves, the temporal and spatial MNLS simulations agree with the statistics observed in laboratory tests, of which the deviation from normal distribution is mainly attributed to the bound wave effects. Finally, considering the peak value of the probability density function is larger than second-order prediction only in the case of long-crested waves, it can be concluded that this departure is mainly generated by the nonlinear instability induced by the quasi-resonant or resonant four-wave interaction.

Under most conditions, the evolution of maximum surface elevation in the temporal model is consistent with that indicated in the spatial model. However, the discrepancy between numerical simulations and laboratory observations is enlarged in the cases of short-crested waves due to different sizes of the observed domain and individual waves, which may also cast some light on the location of extreme events in three-dimensional wave fields considering that these wave probes are limited and only located in the centre of wave basin.

For the selected sea states, the extreme surface elevation is inversely proportional to the initial directional spreading in the three-dimensional wave field, which agrees with the variation tendency of the coefficient of kurtosis. However, both numerical simulations reveal that the maximal extreme surface elevation appears in the case of $N = 840$ rather than in the unidirectional waves.

Additionally, in 3D numerical simulation, a significant evolution of the wave spectrum can be detected over time and space, mainly indicating its width that broadens gradually with various degrees depending on the initial *BFI*. Moreover, a small downshift in the peak frequency can be clearly observed.

6 Modelling Extreme Waves with the HOS Method

6.1 Basic Theory

6.1.1 High-Order Spectral Method

In the case of the irrotational motion of ideal fluid that is considered to be homogeneous, inviscid and incompressible, the flow can be described by a velocity potential $\phi(\boldsymbol{x}, z, t)$, satisfying Laplace's equation within the domain. The governing equation and its boundary conditions are:

$$\nabla^2\phi(\boldsymbol{x}, z, t) = 0, \qquad -h \le z \le \eta(\boldsymbol{x}, t), \tag{6.1}$$

$$\eta_t + \nabla_h\phi \cdot \nabla_h\eta - \phi_z = 0, \qquad z = \eta(\boldsymbol{x}, t), \tag{6.2}$$

$$\phi_t + gz + \frac{1}{2}(\nabla\phi)^2 = 0, \qquad z = \eta(\boldsymbol{x}, t), \tag{6.3}$$

$$\phi_z = 0, \qquad z = -h, \tag{6.4}$$

where $\eta(\boldsymbol{x}, t)$ is the free surface displacement and $\nabla_h \equiv \left(\partial/\partial x, \partial/\partial y\right)$ is the gradient operator in the horizontal (x, y)-plane. The origin of the vertical coordinate z is located at the mean free surface. In terms of the velocity potential $\psi(\boldsymbol{x}, t) = \phi(\boldsymbol{x}, \eta(\boldsymbol{x}, t), t)$ evaluated at the free surface, the kinematic and dynamic boundary conditions can be rewritten as

$$\eta_t + \nabla_h\psi \cdot \nabla_h\eta - W\left[1 + (\nabla_h\eta)^2\right] = 0, \tag{6.5}$$

H. Zhang and C. Guedes Soares, *Numerical Modelling of Extreme Waves*, Synthesis Lectures on Ocean Systems Engineering, https://doi.org/10.1007/978-3-031-77084-5_6

$$\psi_t + g\eta + \frac{1}{2}(\nabla_h\psi)^2 - \frac{1}{2}W^2\left[1 + (\nabla_h\eta)^2\right] = 0, \tag{6.6}$$

where $W(\boldsymbol{x}, t)$ denotes the vertical velocity at the free surface

$$W = \phi_z|_{z=\eta(\boldsymbol{x},t)}. \tag{6.7}$$

In order to follow the temporal evolution of the surface elevation from Eqs. (6.5) and (6.6), a Dirichlet problem of Laplace's equation for $\phi(\boldsymbol{x}, z, t)$ has to be solved first at each time step to obtain W. However, the direct numerical simulations of these equations are rather complex, but they can be treated intelligently, as shown in the high-order spectral method (HOS). The velocity potential $\phi(\boldsymbol{x}, z, t)$ is written in a perturbation series

$$\phi(\boldsymbol{x}, z, t) = \sum_{j=1}^{M} \phi^{(j)}(\boldsymbol{x}, z, t), \tag{6.8}$$

where $\phi^{(j)}$ is assumed to be a quantity of order $O(\varepsilon^j)$ with ε being a measure of wave steepness, and M is the order of approximation of nonlinearity. Applying the Taylor expansion for each $\phi^{(j)}$ around $z = 0$ and collecting terms at each order j, a sequence of boundary conditions for the unknown velocity potential can be obtained:

$$\begin{aligned} \phi^{(1)}(\boldsymbol{x}, 0, t) &= \psi(\boldsymbol{x}, t), \\ \phi^{(j)}(\boldsymbol{x}, 0, t) &= -\sum_{k=1}^{j-1} \frac{\eta^k}{k!} \frac{\partial^k}{\partial z^k} \phi^{(j-k)}(\boldsymbol{x}, 0, t), \quad j = 2, 3, \ldots, M. \end{aligned} \tag{6.9}$$

In addition to Laplace's equation and appropriate bottom boundary condition, these boundary conditions define a series of boundary value problems for $\phi^{(j)}$ in the domain $z \leq 0$. This procedure has simplified the original single Dirichlet problem in a region with a complicated boundary shape $z = \eta$ to a series of M Dirichlet problems in a region with a simple boundary shape $z = 0$.

It is well known that two slightly different versions of the HOS method were simultaneously and independently proposed by West et al. (1987) and Dommermuth and Yue (1987). As summarized in Tanaka (2001a), the procedure to obtain $\phi^{(j)}$ is the same in both versions, but there exists an important difference in the way of calculating W. Dommermuth and Yue computed W from $\phi^{(j)}$ as

$$W(\boldsymbol{x}, t) = -\sum_{j=1}^{M} \sum_{k=0}^{M-j} \frac{\eta^k}{k!} \frac{\partial^{k+1}}{\partial z^{k+1}} \phi^{(j)}(\boldsymbol{x}, 0, t), \tag{6.10}$$

while West et al. first assume a power series in ε for W as

$$W(\boldsymbol{x}, t) = \sum_{j=1}^{M} W^{(j)}, \qquad W^{(j)} = \sum_{k=0}^{j-1} \frac{\eta^k}{k!} \frac{\partial^{k+1}}{\partial z^{k+1}} \phi^{(j-k)}(\boldsymbol{x}, 0, t), \tag{6.11}$$

and then treat all terms in the boundary conditions that contain W in such a way that the consistency of ordering with respect to ε, and hence with respect to $\eta(\boldsymbol{x}, t)$ and $\psi(\boldsymbol{x}, t)$, is retained.

Recently, a comparison of these two methods has been presented by Clamond et al. (2006), whose results reveal that the formulation of Dommermuth and Yue does not converge when the amplitude is very large. In addition, it has been demonstrated in Bonnefoy et al. (2010) that the errors reported in Dommermuth and Yue can be drastically reduced and that convergence properties are conserved up to very high nonlinearity (close to wave breaking limit) by using West form with careful dealiasing. Based on the above discussions, the formulation proposed by West et al. is adopted for the present study (Toffoli et al. 2008b, 2010a; Slunyaev et al. 2014).

To be convenient to set up a numerical wave tank to simulate the laboratory experiment in the following section, the free surface boundary conditions expressed by Eqs. (6.5) and (6.6) are rewritten in the following form:

$$\eta_t - W^{(1)} = F, \tag{6.12}$$

$$\psi_t + g\eta = G, \tag{6.13}$$

where F and G are the nonlinear parts that can be further modified by a new adjustment scheme if the initial condition is complex, such as in the case of JONSWAP sea state (Dommermuth 2000; Ducrozet et al. 2007). In short, the linear part of the equations is analytically integrated while the nonlinear part of the system is computed numerically using the classical fourth-order Runge–Kutta method with an adaptive time step. Note that the high-frequency part of the system is very stiff to solve, but the scheme presented by Fructus et al. (2005) can be used to accelerate the numerical procedure (Ducrozet et al. 2016a).

6.1.2 Breather Solutions of Nonlinear Schrödinger Equation

Even though different numerical methods can solve the Euler equations, the computation is very heavy, and the physics of water waves cannot normally be observed or explained directly. More often than not, a simplification of the equations is needed to get some idea of the interesting phenomenon that may occur in the water wave dynamics. Using the weakly nonlinear approach, the most simple nonlinear Schrödinger (NLS) equation can be derived from the original equations of motion under the hypotheses that the waves are characterized by a small steepness and their spectrum is narrow-banded and the dimensional NLS equation in arbitrary depth is already given in Eq. (3.1).

Comparing with the original equations of motion, the NLS equation presents a lot of advantages. For instance, it is an integrable system; thus, many of its analytical solutions can be explicitly written. In addition to the discovery of the solitons, which are waves that can travel without changing shape, it has been found that the NLS equation has breather solutions such as the Kuznetsov–Ma (Kuznetsov 1977; Ma 1979), the Peregrine (Peregrine 1983) and the Akhmediev (Akhmediev et al. 1985) ones.

For the purpose of simplicity, these breather solutions will be only present in a concise dimensionless form, and the detailed procedure for rescaled variables can be found in the former works (e.g., Chabchoub et al. 2011).

The breather solution proposed by Ma (1979) is periodic in time and has the following dimensionless form of expression for infinite water depth:

$$A_M(X,T) = \frac{\cos(\Omega T - 2i\varphi) - \cosh(\varphi)\cosh(pX)}{\cos(\Omega T) - \cosh(\varphi)\cosh(pX)} \exp(2iT), \tag{6.14}$$

where $\Omega = 2\sinh(2\varphi)$, $p = 2\sinh(\varphi)$ and $\varphi \in \mathbb{R}$. Similarly, Kuznetsov (1977) found another solution, but with an additional term.

Correspondingly, Akhmediev et al. (1985) found another group of breather solutions that is periodic in space with the following dimensionless form of expression for infinite water depth:

$$A_A(X,T) = \frac{\cosh(\Omega T - 2i\varphi) - \cos(\varphi)\cos(pX)}{\cosh(\Omega T) - \cos(\varphi)\cos(pX)} \exp(2iT), \tag{6.15}$$

where $\Omega = 2\sin(2\varphi)$, $p = 2\sin(\varphi)$ and$\varphi \in \mathbb{R}$. If $T \to \pm\infty$, the Akhmediev breather solution tends to a plane wave, which can also appear in the Ma breather solution if $X \to \pm\infty$.

$$\begin{aligned} A_p(X,T) &= \lim_{\varphi\to 0} A_A(X,T) \\ &= \lim_{\varphi\to 0} A_M(X,T) \\ &= \left(1 - \frac{4(1+4iT)}{1+4X^2+16T^2}\right)\exp(2iT). \end{aligned} \tag{6.16}$$

With regard to the Peregrine breather solution (1983) in Eq. (6.16), it can be derived from the limiting case of Ma or Akhmediev breather solution, i.e. $\varphi \to 0$, when the period of either solution becomes infinite.

To better understand the characteristics of these solutions, some examples are given in Fig. 6.1. Generally speaking, the breather solutions start in space or time with an almost constant background (corresponding to a quasi-monochromatic wave in terms of the surface elevation). Then, during their evolutions, the perturbation on the envelope

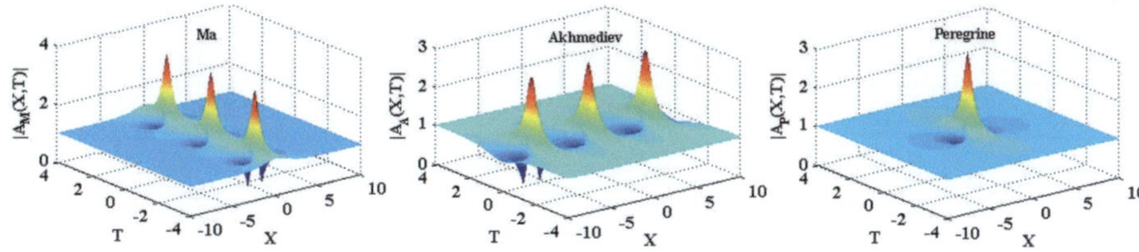

Fig. 6.1 Sketches of the three types of breather solutions

grows with different growth rates and finally reaches a much larger amplitude than the initial one. Therefore, all three types of solutions can be considered as the prototypes of rogue waves.

However, the Peregrine solution differs from the other two types of solutions with a peculiarity that appears only once, i.e., the extreme event is localized in space and time. The periodicity can be observed neither in space nor in time. The Peregrine solution is characterized by an amplification factor of three independent of the water depth (Onorato et al. 2013b), meaning that the maximal amplitude can always reach three times the initial one. Moreover, the confirmation of laboratory observation has been reported for the first time in the works of Chabchoub et al. (2012a, 2012b).

6.2 Experimental and Numerical Setup

6.2.1 Wave Basin and Ma Breather Solution

In this work, the Ma breather solution is studied in a series of laboratory experiments performed in the seakeeping basin of the Ocean Engineering Division of Technical University Berlin. Detailed descriptions of the facilities and experiments can be found in the earlier work (Clauss et al. 2011). Here, only a concise introduction is provided.

Mechanically generated waves are produced in a large rectangular wave basin with dimensions of 110 m × 8 m. On the one hand, an electrically driven piston-type wave generator is installed and fully controlled by the computer, which can generate regular waves, transient wave packages, deterministic irregular sea states with defined characteristics, and tailored critical wave sequences. On the opposite side, it is equipped with an absorbing sloping beach that suppresses the effect of wave reflection. The water depth is set to be 1 m for the present experiments. Ten surface piercing resistance-type wave probes are installed downstream of the wave basin, starting at 15 m and ending at 60 m with an interval of 5 m between each gauge.

The main parameters of the ten investigated cases are presented in Table 6.1, where the second column shows the carrier wave periods and the third column presents the

Table 6.1 Parameters on the investigated Ma breather solution

Case	T_0 (s)	ε	k_0h	*BFI*
1	1.25	0.114	2.58	1.48
2	1.25	0.136	2.58	1.85
3	1.25	0.147	2.58	2.01
4	1.4	0.114	2.05	1.27
5	1.4	0.147	2.05	1.66
6	1.4	0.170	2.05	1.91
7	1.5	0.114	1.79	1.11
8	1.5	0.147	1.79	1.45
9	1.5	0.170	1.79	1.67
10	1.5	0.182	1.79	1.78

associated wave steepness $\varepsilon = k_0 a_0$, of which different values have been assigned for the same carrier wave period to indicate the influence of wave steepness.

Different lengths of carrier waves have been chosen to give various dimensionless water depths, including the effect of water depth, as presented in the fourth column. Following Onorato et al. (2001), the Benjamin–Feir Index (*BFI*), is introduced in the last column to take into account the effects of steepness and water depth simultaneously.

6.2.2 Numerical Schemes

In the following numerical simulation, the approximately fully nonlinear HOS model will be used to simulate the laboratory observation. Since only point measurements of the surface elevation, a function of time at the ten different locations, are available, it is not straightforward to initialize the HOS model, which requires the initial surface elevation and the velocity potential in the entire computational domain.

To overcome this problem, the method proposed by Goullet and Choi (2011) is adopted here. If $\eta_1(t)$ denotes a time history of the surface elevation measured at the first probe, which without loss of generality is assumed to be located at $x = 0$, the surface elevation and the velocity potential can be expressed as a linear superposition of sinusoidal waves propagating in the x-direction

$$\eta(x, t) = \sum_n a_n \exp[i(k_n x - \omega_n t)] + c.c., \tag{6.17}$$

$$\psi(x, t) = \sum_n c_n \exp[i(k_n x - \omega_n t)] + c.c., \tag{6.18}$$

where $\omega_n = 2\pi n/T_1$ with T_1 being the total time interval for $\eta_1(t)$ and the wave number k_n are computed using the linear dispersion relation, $k_n = \omega_n^2 / \left[g \tanh(k_n h)\right]$ with an iterative method.

The coefficients a_n in Eq. (6.17) with $x = 0$ are found from the Fourier transform of $\eta_1(t)$, i.e., $\eta_1(t) = \sum_n a_n \exp(-i\omega_n t)$. On the other hand, c_n is simply obtained from the linear relationship between η and ψ on the mean water surface, which yields $c_n = -ia_n\omega_n / k_n$.

The HOS model can be initialized by evaluating Eqs. (6.17) and (6.18) at $t = 0$. To be more consistent with reality, the computational domain is decomposed into two regions: the linear and nonlinear parts. In the upstream region of the first probe, the linearized Eqs. (6.12) and (6.13) with $F = G = 0$ are solved so that the waves reaching the first probe match exactly with the experimental observations, while the nonlinear Eqs. (6.5) and (6.6) with $M > 1$ are solved in the downstream region of the first probe (Ducrozet et al. 2007; Goullet and Choi 2011).

This requirement can be implemented by multiplying the nonlinear terms in evolution Eqs. (6.12) and (6.13) by a smooth Heaviside-like function varying from 0 to 1. The width of the transition region centred at the first probe is two grid points.

Downstream the numerical wave tank, the surface elevations derived from the HOS method are recorded at some specified locations as a set of time series for direct comparison with laboratory observations. An alternative to this procedure is to use a 'real' numerical wave tank (i.e., with a wavemaker, an absorbing zone, etc.) to reproduce the experiments (Ducrozet et al. 2012).

It has to be pointed out that this initialization scheme is constructed according to the linear relationship between the free surface elevation and the velocity potential. Hence, it may be inaccurate, and the error could result in some discrepancies between numerical simulations and experimental measurements.

However, as shown in the later comparison, the discrepancy is insignificant. Even though it is for intermediate water depth in this study, the fact that the linear velocity potential is valid up to the third order in the Stokes expansion of deepwater waves may cast some light on this point.

6.3 Experimental and Numerical Results

6.3.1 Experimental Results

The variations of maximum wave crest and maximum wave height in the Ma breather solution are presented in Figs. 6.2 and 6.3, respectively, as a function of the distance to the wavemaker, which has been scaled by the carrier wavelength L_0 that is corresponding to the initial carrier wave period T_0 with a linear dispersion relationship.

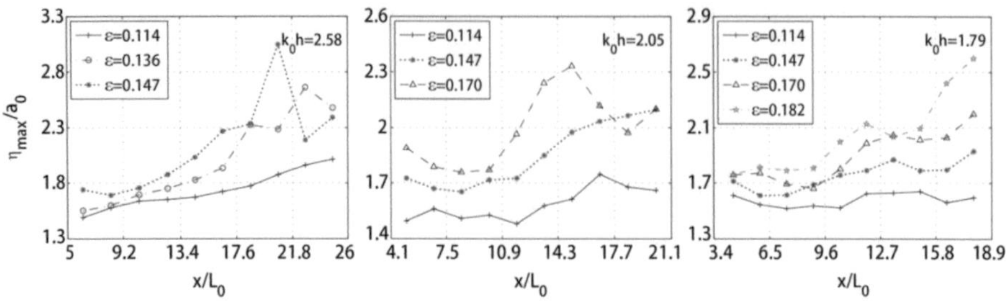

Fig. 6.2 Spatial variation of maximum wave crest in the cases with different steepness and relative water depth

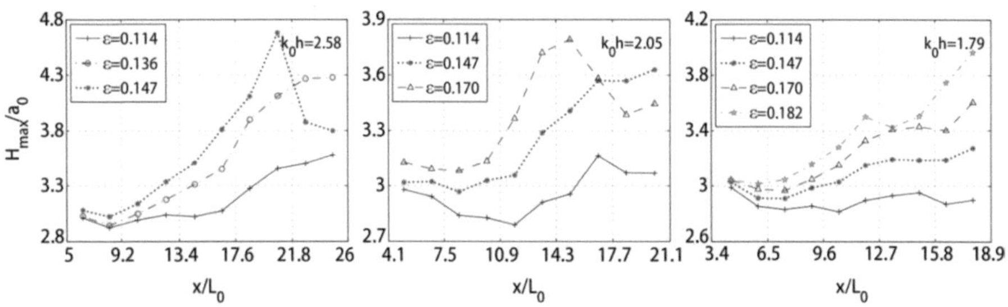

Fig. 6.3 Spatial variation of maximum wave height in the cases with different steepness and relative water depth

Apparently, for the same test case, both figures present a similar variation tendency, meaning that the generation of maximum wave height is not only generally accompanied by but also mainly composed of a very large wave crest.

Moreover, the extreme wave height in Fig. 6.3 appears in the case of $\varepsilon = 0.147$ with $k_0h = 2.58$, which corresponds to the maximum $BFI = 2.01$ in Table 6.1. It is about 4.7 times the initial amplitude a_0 in the location of $20.7L_0$ (Gauge 8), the distance of which to the wavemaker is in agreement with the observation in JONSWAP sea state, i.e., a range from $15L_0$ to $30L_0$ depending on the initial BFI (Onorato et al. 2006; Zhang et al. 2014, 2015).

As expected, the maximum wave height and growth rate are proportional to the initial wave steepness for the same water depth. Suppose the initial nonlinearity represented by the wave steepness of the breather solutions is the same. In that case, the amplitude of the maximum wave that appeared at each location will be largely reduced with the decreasing water depth.

Moreover, the distance that the group has to travel to reach the extreme wave height will increase significantly, approaching infinity if k_0h is close to 1.36, which is consistent with the features of the NLS equation and shown by Clauss et al. (2011).

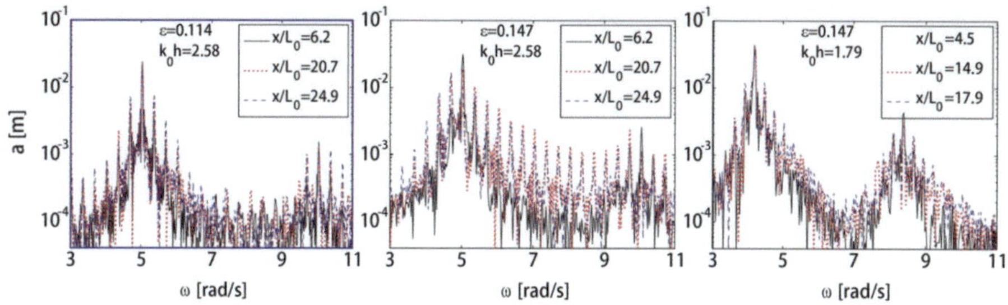

Fig. 6.4 Wave amplitude spectra measured at Gauges 1, 8 and 10 along the tank for cases 1 (left), 3 (middle) and 8 (right), respectively

To better reveal the influence of wave steepness and water depth on the evolution of Ma breather solution, the wave spectra of three typical cases (1, 3, and 8) listed in Table 6.1 have been compared in Fig. 6.4. Due to the limitation of the space, only the amplitude spectra of wave series measured at three locations (Gauges 1, 8 and 10) scaled by the carrier wavelength L_0 are presented here.

Apparently, if the initial wave steepness is small (e.g., $\varepsilon = 0.114$ in case 1) or the water is shallow (e.g., $k_0 h = 1.79$ in case 8), the wave spectrum will not significantly change the evolutionary process. Otherwise, the variation of the wave spectrum will be complex, as shown in case 3 of Fig. 6.4, where the energy of the carrier wave decreases sharply and the peak frequency downshifts to the closest lower harmonic component.

The other notable change is reflected in the range $\omega_p < \omega < 2\omega_p$. The magnitudes of these wave components increase quickly before the location $20.7L_0$ where the extreme wave height is observed. After this position, part of the energy in the range $\omega_p < \omega < 2\omega_p$ downshifts to the lower frequency section, i.e. $0 < \omega < \omega_p$. A similar variation is also observed in laboratory experiments with Peregrine breather solution (Shemer and Alperovich 2013).

Furthermore, it seems that the measuring section in the seakeeping basin is not long enough to evaluate the evolution of the Ma breather solution. Besides, the inherent mechanics cannot be indicated only by laboratory experiments. Thus, the numerical simulation performed with the HOS method becomes mandatory in this research.

6.3.2 Numerical Simulation of Ma Breather Solution

Before studying the breather solutions with the numerical method, it is necessary to check the capability of HOS model in simulating the laboratory observations.

As shown in Fig. 6.5, if the initial *BFI* is small which is mainly caused by the insignificant nonlinearity (case 1), the numerical simulation can present a good agreement with the experimental measurement in all locations.

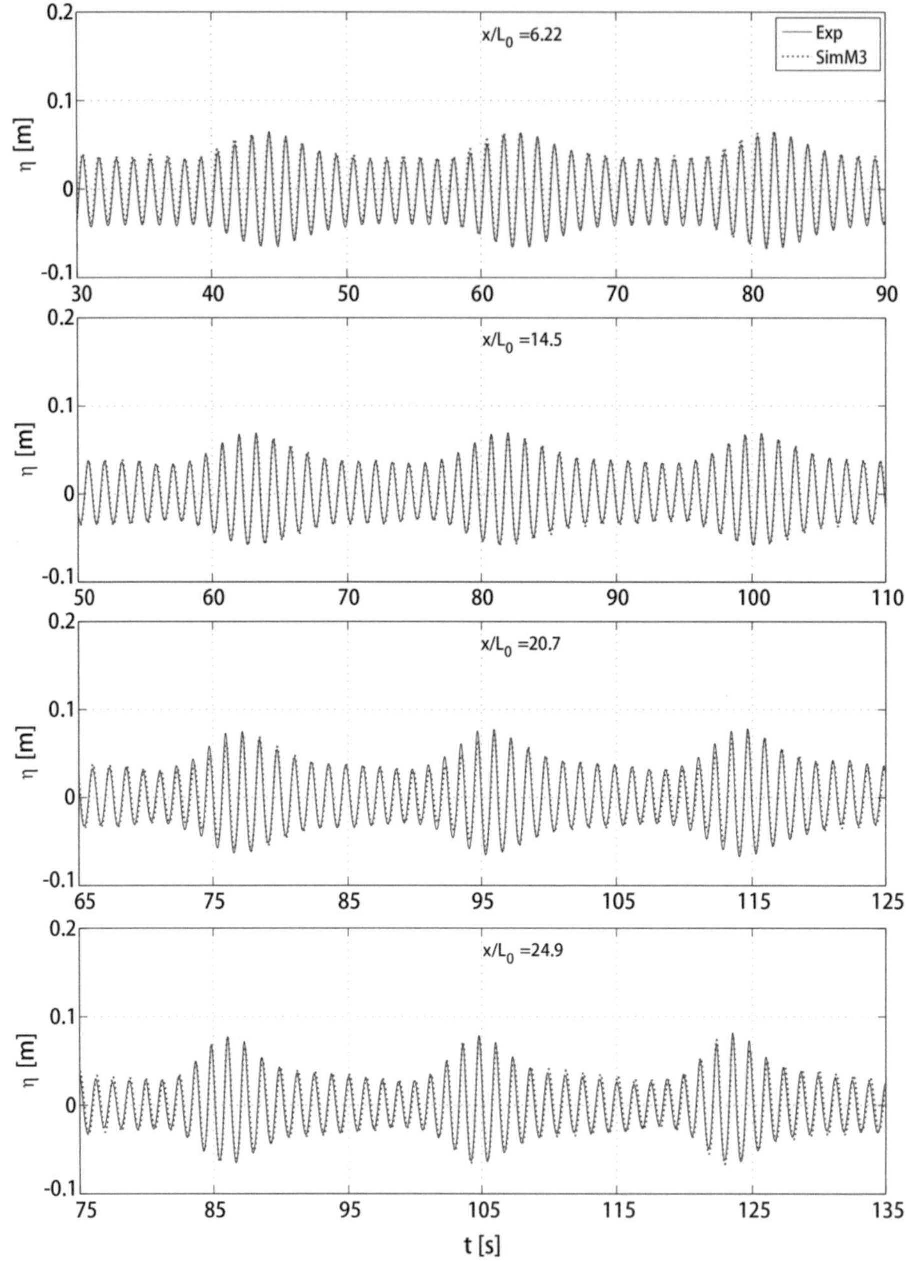

Fig. 6.5 Numerical simulation of the laboratory experiment in case 1. The locations corresponding to the Gauges 1, 5, 8 and 10 have been scaled by the carrier wave length L_0. 'Exp' and 'SimM3' represent the laboratory observation and numerical simulation with $M = 3$, respectively

If the initial *BFI* is very large, such as case 3 presented in Fig. 6.6, the HOS method can perform very well before the location of extreme wave generation ($x = 20.7L_0$) where the wave breaking is observed in the wave basin. After this location, the numerical simulation is still reliable even though some discrepancies can be observed due to the severe dissipation of wave energy induced by wave breaking in the experiment.

It must be pointed out that a small phase shifting that appeared in the numerical simulation has to be removed to get a better comparison. On the whole, it can be argued that the HOS method has the ability to accurately reproduce the evolutionary process of breather solutions, particularly in the case with small initial *BFI*.

The above numerical simulations are carried out with $M = 3$, which can include the quasi-resonant four-wave interaction process in the HOS method and is strictly equivalent to Zakharov's equation (Tanaka 2001b). Numerical tests in the simulation also indicate that a further increase of the M value does not significantly improve the agreement with laboratory observation except for the formidable computational time evolving as M^2 (Ducrozet et al. 2016a).

Since the evolutionary process in the cases with small *BFI* is usually very slow but similar to the case with large *BFI*, to be concise, only one wave group in case 3 is further studied herein with different orders of nonlinearity.

It is well known that if $M = 1$, the HOS model's wave dynamics will be linear. As demonstrated by the red dot line in Fig. 6.7, the same initial wave group in the Ma breather solution will be almost constant in the whole evolution process under this condition because no nonlinear effect is involved. If $M = 2$, meaning that the nonlinearity contained in the numerical model is up to the second order, the shape of the same initial wave group will be changed significantly but still not enough to form the original extreme wave that appeared in the location of $20.7L_0$.

Alternatively, it can be explained that due to the higher-order nonlinear effects, the back large amplitude waves travel much faster than the front small amplitude waves, eventually leading to an intensive accumulation of energy on a specific front small wave.

Now based on the discussions of Figs. 6.6 and 6.7, it reveals that it is the quasi-resonant four-wave interaction that sucks energy from the following waves to create the enormous amplitude in the Ma breather solution. To further see the inherent variation of the wave group at a much longer time scale, the corresponding wave number spectrum in numerical simulation has been directly plotted in Fig. 6.8.

Evidently, for the linear evolutionary process such as $M = 1$, the wave components do not change in the breather solution. If the nonlinearity is increased up to the second order, the amplitude of the carrier wave will decrease sharply with the peak frequency downshifting to the subharmonic component. Suppose the nonlinearity continues to be increased, in addition to the similar phenomena detected in the case of $\mathrm{M} = 2$. In that case, the wave number spectrum will approach a continuous one with the peak frequency further downshifted and the most energy relocated in the lower wave number section.

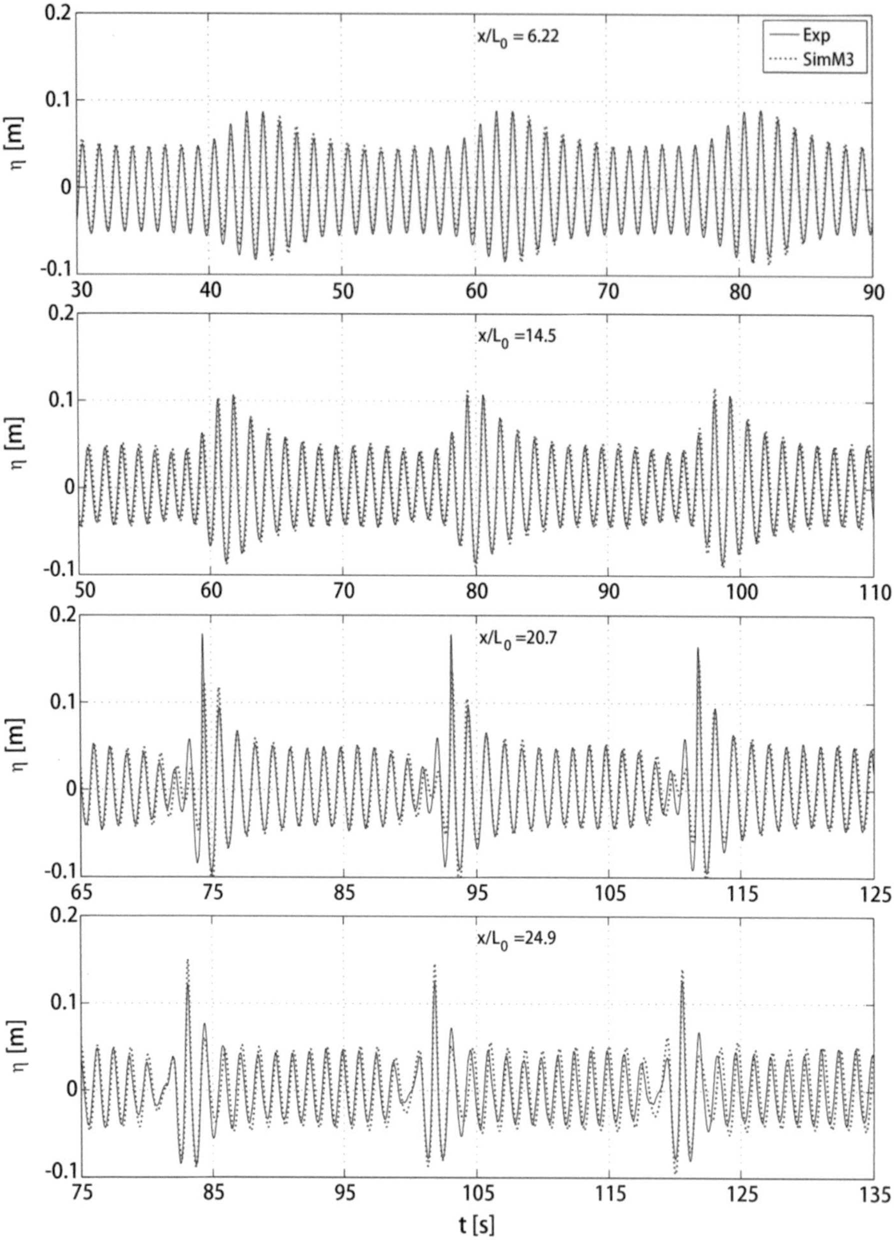

Fig. 6.6 Numerical simulation of the laboratory experiment in case 3. The locations and legends have the same meanings as those in Fig. 6.5

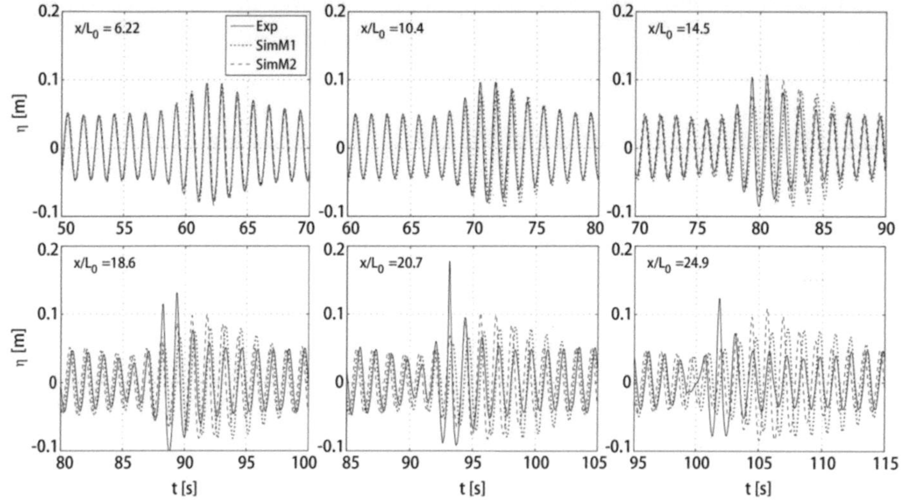

Fig. 6.7 Comparison between laboratory observation and numerical simulations of case 3 with different orders of nonlinearity. The locations corresponding to the Gauges 1, 3, 5, 7, 8 and 10, are scaled by the carrier wave length. 'Exp', 'Sim1' and 'Sim2' represent laboratory observation and numerical simulations with $M = 1$ and $M = 2$, respectively

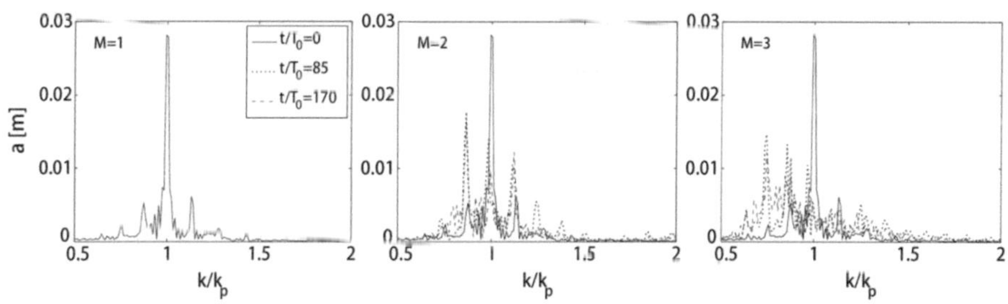

Fig. 6.8 Longer-term variation of wave number spectrum of simulated wave series in case 3 with different orders of nonlinearity

A similar variation has been observed for a Stokes wave disturbed by unstable modes in the work of Chalikov (2007), where the fully nonlinear CS model is applied. The final form of the wave spectrum indicates that the Ma breather solution cannot return to its initial shape after breathing and will finally evolve into an irregular wave series.

Apparently, it is beyond the validity of the NLS equation due to the complex nonlinear effects and the wave breaking in reality.

Finally, it has to be mentioned that some laboratory experiments reveal that wave breaking can lead to a downshift in peak frequency. However, the numerical simulations (no breaking) reveal that the third-order nonlinear effect or modulational instability results in the frequency downshifting. Consequently, it may be argued that wave breaking is just an accompanying phenomenon rather than a fundamental reason.

6.3.3 Numerical Simulation of Peregrine Breather Solution

As mentioned before, the Peregrine breather solution in Eq. (6.19) is a limited case of Ma breather solution and presents many unique features, such as an amplification factor of three and only breathing once in time and space.

$$A_p(x,t) = a_0 \exp\left(-\frac{ik_0^2 a_0^2 \omega_0}{2} t\right)\left(1 - \frac{4\left(1 - ik_0^2 a_0^2 \omega_0 t\right)}{1 + \left[2\sqrt{2}k_0^2 a_0\left(x - C_g t\right)\right]^2 + k_0^4 a_0^4 \omega_0^2 t^2}\right). \tag{6.19}$$

where C_g is the wave group velocity. After specifying the values of parameters (a_0 and k_0) of the initial carrier wave, it is straightforward to derive the theoretical surface elevation to the first order with the following equation:

$$\eta_P = \mathrm{Re}\{A_P(x,t)\exp[i(k_0 x - \omega_0 t)]\}. \tag{6.20}$$

The observation of this solution for water waves was first reported by Chabchoub et al. (2012a) in a wave tank, and a similar experiment was carried out by Shemer and Alperovich (2013) who found that the location of extreme crest locates in $x = 11.6$ m rather than $x = 9.1$ m. Moreover, the extreme crest height is 3.17 times the initial magnitude, larger than the three predicted by the theoretical breather solution.

In this section, the same laboratory experiment is simulated using the HOS method to explore its features further. As represented by the red solid line (Case1) in Fig. 6.9, the initial surface elevation at $x = 0$ m is obtained directly from Eq. (6.19), of which the detailed parameter values can be found in the work of Chabchoub et al. (2011). As expected, the maximal wave crest appears in the location $x = 9.1$ m (see the second picture) in the numerical simulation, very similar to Fig. 5 in their work.

If a phase angle is added to Eq. (6.19), the initial surface elevation plotted by the blue dot line (Case 2) will be slightly different from the exact one. Such kind of difference will lead to a longer distance ($x = 14.8$ m) for the wave group to travel to achieve the first extreme wave crest, in addition to a larger amplification factor that is more than three, as illustrated in the third picture of Fig. 6.9. Now it can be concluded that the initial phase angle can play a significant role on the evolutionary property of breather solution.

With regard to the variations of the amplitude spectra along the numerical wave tank, the same evolutionary tendencies can be detected as those described by Shemer and Alperovich (2013) in the laboratory experiments. Moreover, the long-term variation trend is similar to that in Fig. 6.8 in the Ma breather solution and is not repeated here.

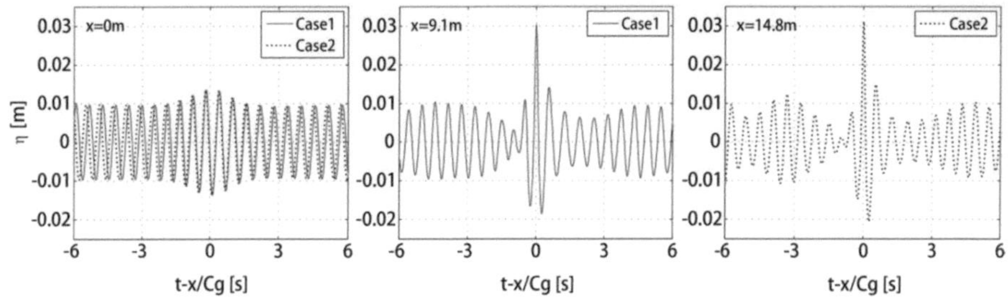

Fig. 6.9 Numerical simulation of Peregrine breather solution with different initial phase angles. For both cases, numerical simulations are carried out with $M = 3$

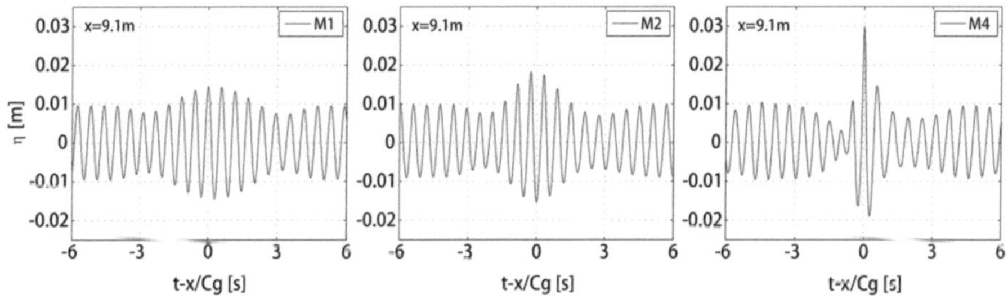

Fig. 6.10 Numerical simulation of Peregrine breather solution with different orders of nonlinearity. M1, M2 and M4 represent $M = 1$, $M = 2$ and $M = 4$ in the numerical simulations, respectively

In Fig. 6.10, three different surface elevations at $x = 9.1$ m are presented for comparison. They have the same initial surface elevation but are governed by the HOS model with different orders of nonlinearity. Obviously, conclusions similar to those in the Ma breather solution can be obtained. It is confirmed again that the formation of a very large wave that appeared in various breather solutions is actually due to modulational instability.

It has to be admitted that the discrepancy between numerical simulation and laboratory observation (see Fig. 5 in the work of Chabchoub et al. 2011) in the location $x = 9.1$ m cannot be avoided because the total energy in the laboratory experiment will be dissipated by the wave breaking in the main form of white capping as well as a little sidewall friction at times while it will be almost constant in the numerical simulation.

6.4 Conclusion

The laboratory tests found that the generation of maximum wave height that appeared in the Ma breather solution is usually accompanied by a very large crest amplitude. Comparisons of the ten cases of Ma breather solutions indicate that the maximal wave height and its growth rate are proportional to the initial wave steepness in the same dimensionless water depth.

If the wave series have the same initial condition, the maximum wave height at each location can be largely reduced by the decreasing water depth and the evolutionary process will become very slow. Meanwhile, the wave spectrum has a significant change in the case of a larger initial *BFI*. The amplitude of the carrier wave decreases sharply, and peak frequency downshifts can be observed.

Besides, the magnitudes of wave components in the range $\omega_p < \omega < 2\omega_p$ first increase quickly before the location of generation of an extreme wave. Then, part of the energy in this range is downshifted to the lower frequency section, i.e. $0 < \omega < \omega_p$.

In the numerical study, the evolutions of breather solutions with the same initial wave groups are governed by the HOS model with different orders of nonlinearity. If $M = 1$, the evolutionary process is linear without a significant change. If $M = 2$, the nonlinearity is still insufficient to form the original extreme wave that appeared in both Ma and Peregrine breather solutions.

Only the quasi-resonant four-wave interaction or modulational instability is fully included can the simulation be consistent with the laboratory observations. Suppose the evolutionary time scale is long enough. In that case, the wave spectrum will eventually approach a continuous one with the peak frequency downshifted and the most energy relocated in the lower frequency domain.

Numerical experiments also reveal that a small initial phase shift in the Peregrine solution can lead to a longer distance for the wave group to travel to achieve the extreme amplitude and a larger amplification factor than that predicted by the theoretical solution.

7 Modelling Extreme Waves with the Chalikov–Sheinin Model

7.1 Basic Theory

7.1.1 Chalikov–Sheinin Model

It is well known that in two-dimensional space, the motion of ideal fluid can be described by the following governing equation and its boundary conditions:

$$\nabla^2\phi(x, z, t) = 0, \qquad -h \le z \le \eta(x, t), \tag{7.1}$$

$$\eta_t + \phi_x\eta_x - \phi_z = 0, \qquad z = \eta(x, t), \tag{7.2}$$

$$\phi_t + z + \frac{1}{2}(\nabla\phi)^2 = 0, \qquad z = \eta(x, t), \tag{7.3}$$

$$\phi_z = 0, \qquad z = -h, \tag{7.4}$$

where all variables still have the same meaning as before but are already scaled to guarantee that the horizontal spatial domain in computation is equal to 2π (Chalikov and Sheinin 1998, 2005). The origin of the vertical coordinate z is located at the mean free surface.

If the 2D water wave, described by the principal potential Eqs. (7.1)–(7.4), is periodic in time and horizontal space, a more accurate numerical method can be used to solve these equations. Due to the periodicity condition, the conformal mapping can be represented by the Fourier series:

$$x = \xi + x_0(\tau) + \sum_{-M \le k \le M, k \ne 0} \eta_{-k}(\tau)\frac{\cosh k(\zeta + H)}{\sinh kH}\vartheta_k(\xi), \tag{7.5}$$

H. Zhang and C. Guedes Soares, *Numerical Modelling of Extreme Waves*, Synthesis Lectures on Ocean Systems Engineering, https://doi.org/10.1007/978-3-031-77084-5_7

$$z = \zeta + \eta_0(\tau) + \sum_{-M \le k \le M, k \ne 0} \eta_k(\tau) \frac{\sin hk(\zeta + H)}{\sinh kH} \vartheta_k(\xi), \tag{7.6}$$

where x and z are the Cartesian coordinates, while ξ and ζ are the conformal surface-following coordinates. In the new coordinate system, denoting time by τ, H represents the water depth depending on τ and η_k are coefficients of Fourier expansion of η (ξ, τ) with respect to the new horizontal coordinate ξ.

$$\hat{\eta}(\xi, \tau) = \eta(x(\xi, \zeta = 0, \tau), t = \tau) = \sum_{-M \le k \le M} \eta_k(\tau)\ \vartheta_k(\xi), \tag{7.7}$$

where ϑ_k denotes the function

$$\vartheta_k(\xi) = \begin{cases} \cos k\xi, k \ge 0 \\ \sin k\xi, k < 0 \end{cases}, \tag{7.8}$$

and M is the truncation number.

The representation of Fourier transform with definition (7.8) is nontraditional but more convenient for calculations with real numbers $\sum (A_k \vartheta_k)_\xi = -\sum kA_{-k}\vartheta_k$.

Note that the definitions of coordinates ξ and ζ are expressed on the Fourier coefficients of surface elevation. It then follows from Eqs. (7.5) and (7.6) that time derivatives x_τ and z_τ are connected by Cauchy–Riemann relations.

In the new (ξ, ζ) coordinate, the Laplace equation still retains its form due to conformity, while the kinematic and dynamic boundary equations have been changed:

$$\Phi_{\xi\xi} + \Phi_{\zeta\zeta} = 0, \tag{7.9}$$

$$z_\tau = -x_\xi \zeta_t - z_\xi \xi_t, \tag{7.10}$$

$$\varphi_\tau = -\xi_t \varphi_\xi - \frac{1}{2} J^{-1} \left(\varphi_\xi^2 - \Phi_\zeta^2\right) - z, \tag{7.11}$$

where Eqs. (7.10) and (7.11) are written for the surface $= \zeta\ 0$ (so that $z = \hat{\eta}$, as represented by Eq. (7.7)), and J is the Jacobian of the transformation

$$J = x_\xi^2 + z_\xi^2 = x_\zeta^2 + z_\zeta^2. \tag{7.12}$$

In addition,

$$\zeta_t = -\left(J^{-1}\Phi_\zeta\right)_{\zeta=0}, \tag{7.13}$$

and $\varphi = \Phi(\zeta = 0)$, denotes the velocity potential on the surface $z = \hat{\eta}$. Actually, ξ_t is a generalization of the Hilbert transform of ζ_t, which for $k \ne 0$ may be defined in Fourier space as

$$(\xi_t)_k = (\zeta_t)_{-k} cothkH. \tag{7.14}$$

Finally, the new form of the bottom boundary condition, which assumes the vanishing of vertical velocity at the bottom, is given

$$\Phi_\zeta(\xi, \zeta = -H, \tau) = 0. \tag{7.15}$$

The solution of the Laplace Eq. (7.9) with boundary condition (7.15) readily yields the following Fourier expansion, which reduces the system (7.9)–(7.11) to a 2D problem:

$$\Phi = \sum_{-M \le k \le M} \phi_k(\tau) \frac{\cos hk(\zeta + H)}{coshkH} \vartheta_k(\xi), \tag{7.16}$$

where ϕ_k are Fourier coefficients of the surface potential φ. In this way, Eqs. (7.10), (7.11), (7.13) and (7.14) constitute a closed system of prognostic equations for the surface functions $z(\xi, \zeta = 0, \tau) = \eta(\xi, \tau)$ and $\Phi(\xi, \zeta = 0, \tau)$. It may also be regarded as a system of ordinary differential equations for the Fourier coefficients η_k and ϕ_k. Its explicit form would be somewhat complicated but not necessary.

For the deepwater condition, i.e. $H \to \infty$, the coefficients in expansions (7.5), (7.6) and (7.16) can become much simpler. As described in the following section, the Fourier transform method will be used to calculate the nonlinearities. Indeed, the transform method may be called an exact method if the order of nonlinearity is finite. Otherwise, the accuracy of the method should be established empirically.

Note that the above equations have been written in dimensionless forms with the following scales: length L_1, where $2\pi L_1$ is the dimensional length in the horizontal direction, time $L_1^{1/2} g^{-1/2}$ and velocity potential $L_1^{3/2} g^{-1/2}$ (g is the acceleration of gravity). This study does not consider capillary effects and external pressure since they are not very important in simulating laboratory-generated extreme waves (Zhang et al. 2017b).

7.1.2 Numerical Schemes

For a spatial approximation of the system (7.10) and (7.11), a Galerkin-type or spectral method is used on the basis of Fourier expansion of the prognostic variables with a finite truncation number M. The system is thus reduced to that of ordinary differential equations with $4M + 2$ Fourier coefficients $\eta_k(\tau)$, $\phi_k(\tau)$, $-M \le k \le M$:

$$\dot{\eta}_k = E_k(\eta_{-M}, \eta_{-M+1}, \ldots, \eta_M, \phi_{-M}, \phi_{-M+1}, \ldots, \phi_M), \tag{7.17}$$

$$\dot{\phi}_k = F_k(\eta_{-M}, \eta_{-M+1}, \ldots, \eta_M, \phi_{-M}, \phi_{-M+1}, \ldots, \phi_M), \tag{7.18}$$

where E_k and F_k are respectively the Fourier expansion coefficients for the right-hand sides of Eqs. (7.10) and (7.11) as functions of ξ.

To calculate E_k and F_k as functions of the prognostic variables η_k and ϕ_k, differentiation of the Fourier series is used (the spatial derivatives are thus evaluated exactly) and the nonlinearities are calculated with the so-called transform method by their evaluation on a spatial grid.

For example, if $Y(u(\xi), v(\xi), w(\xi), ...)$ is a nonlinear function, of which the arguments are represented by their own Fourier expansions, grid-point values $u(\xi^{(j)})$,$v(\xi^{(j)})$ $w(\xi^{(j)})$, ... are first calculated (in other words, inverse Fourier transforms are performed).

Then $Y^{(j)} = Y\left(u(\xi^{(j)}), v(\xi^{(j)}), w(\xi^{(j)}), ...\right)$ can be evaluated at each grid point. Finally, the Fourier coefficients Y_k of the function Y are obtained by a direct Fourier transform. Here $\xi^{(j)} = 2\pi(j-1)/N$ and N is the number of grid points.

For the method to be a purely Galerkin one, that is, to ensure the minimum mean square approximation error, the Fourier coefficients E_k and F_k must be evaluated exactly for $-M \le k \le M$. Due to this reason, one must choose

$$N > (v_1 + 1)M, \tag{7.19}$$

where v_1 is the maximum order of nonlinearity. Since the right-hand sides of Eqs. (7.10) and (7.11) include division with Jacobian J, the nonlinearity is of infinite order such that the above condition on N cannot be met.

However, numerical integrations show that if one chooses a value of N to ensure the exact evaluation of cubic nonlinearities ($v_1 = 3$ in Eq. (7.19)), a further increase in N (with a fixed M) does not affect the numerical solution. For the results presented in this chapter, $N = 4M$ was chosen.

No matter how high the spectral resolution, simulations of strongly nonlinear waves require that the energy flux be parameterized into several parts of the spectrum. If ignored, spurious energy accumulations at large wave numbers can corrupt the numerical solution. Simple dissipation terms are therefore added to the right-hand sides of Eqs. (7.17) and (7.18):

$$\dot{\eta}_k = E_k - \mu_k \eta_k, \tag{7.20}$$

$$\dot{\phi}_k = F_k - \mu_k \phi_k, \tag{7.21}$$

with

$$\mu_k = \begin{cases} rM\left(\dfrac{|k| - k_d}{M - k_d}\right)^2 & if\,|k| > k_d \\ 0 & otherwise \end{cases}, \tag{7.22}$$

where $k_d = M/2$ and $r = 0.25$. The dissipation terms effectively absorb energy at wave numbers close to the truncation number M while leaving longer waves virtually intact. With regard to the time integration, the fourth-order Runge–Kutta scheme is used. The choice of time step was done empirically and set equal to 0.002 in the numerical simulation of extreme waves in Sect. 7.3.

To simulate the nonlinear water waves characterized by the JONSWAP wave spectrum, the initial condition for the Fourier coefficients of surface elevation $\eta(x)$ is assigned in the form

$$\begin{cases} \eta_i &= \sqrt{2S_i(K)\Delta K}\cos\theta_i \\ \eta_{-i} &= -\sqrt{2S_i(K)\Delta K}\sin\theta_i \\ i &= 1, 2, 3, ..., k_m \end{cases}, \tag{7.23}$$

where η_i and η_{-i} are the Fourier coefficients in the Cartesian coordinates, and θ is the random (over i and over different runs) phases of modes uniformly distributed in the interval $[0, 2\pi)$. Here $S(K)$ is the wave number spectrum with the following expression:

$$S(K) = \frac{\alpha}{2K^3}\exp\left[-\frac{5}{4}\left(\frac{K_p}{K}\right)^2\right]\gamma^{\exp\left[-\left(\sqrt{K/K_p}-1\right)^2/\left(2\sigma_0^2\right)\right]}, \tag{7.24}$$

where K_p is the peak wavenumber and the parameter σ_0 has the standard values: 0.07 for $K \le K_p$ and 0.09 for $K > K_p$. The Philips parameter α is related to the desired nondimensional significant wave height, and the peak enhancement factor γ specifies the spectral bandwidth.

Using the linear wave theory, the other initial condition for the Fourier coefficients of velocity potential at the free surface can be obtained easily by the following relation:

$$\phi_k = \frac{\text{sign}(k)\eta_{-k}}{\sqrt{k\tanh(kH)}}. \tag{7.25}$$

Note that the initial data are usually given in the Cartesian coordinates, which must be converted into the (ξ, ζ) coordinates. For this purpose and postprocessing of the results, an interpolating algorithm based on the periodic spline functions has been developed, which can carry out the transformation with high accuracy.

At last, wave breaking is regarded to happen in the numerical simulation if the water surface becomes vertical at any point. The criterion for terminating the model run is defined by the first appearance of a non-single value of surface η:

$$x(j+1) < x(j), j = 1, 2, 3, ..., N-1. \tag{7.26}$$

7.2 Comparison Between HOS Method and Chalikov–Sheinin Model

7.2.1 Capability of the Chalikov–Sheinin Model

In the work of West et al. (1987), two single sine waves with initial slopes 0.3142 and 0.4712 were simulated with the HOS method, showing that the first wave steepens and fluctuates in form without breaking, while the second breaks in a short time with the last moment sketched in Fig. 7.1a. Meanwhile, a single cosine wave with an initial slope $\varepsilon = 0.4$ is also presented in Fig. 7.1b, governed by the Chalikov-Sheinin (CS) model, propagating to the onset of wave breaking. When the overturning begins, the crest sharpens dramatically, accompanied by a flattening of the trough, and the maximal velocity at the crest exceeds the linear phase velocity by 1.4 times.

Comparing with the wave breaking shape presented in the HOS method in Fig. 7.1a, it demonstrates one of the reasons why the conformal mapping method is used in the CS model, that is, the ability to reproduce wave surface profiles that are multi-valued functions of the horizontal coordinate. Essentially, conformal mapping exists up to the moment when an overturning volume of water touches the surface. However, the number of Fourier modes needed increases very quickly in such an evolution, and the calculations will terminate much earlier due to the strong crest instability, if some special measures such as smoothing are not taken.

Moreover, Chalikov and Sheinin (2005) reported that for an initially monochromatic wave, the critical value of the initial slope for wave breaking lies between 0.27 and 0.28, which is far below the range between 0.3142 and 0.4712 given by HOS method. Even though the resolution scheme and dissipation terms can play a role in determining

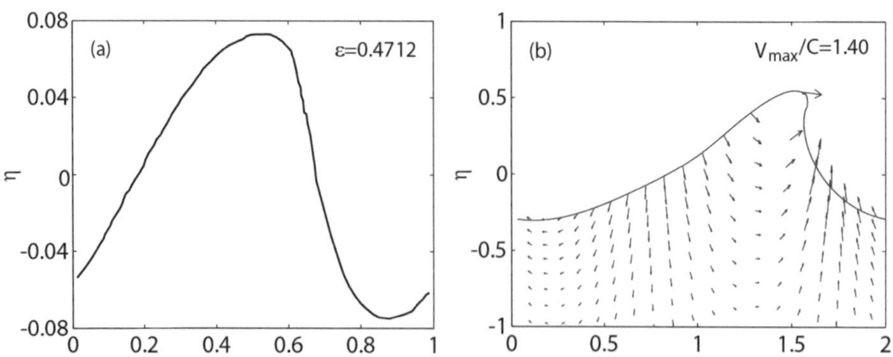

Fig. 7.1 **a** From West et al. (1987); **b** surface profiles, velocity vector fields (scaled by the linear phase velocity of the base wave) at the onset of wave breaking for the initially monochromatic wave with the expression $\eta = 0.4\cos(x)$

the critical slope of wave breaking, the lower threshold in the CS model is reasonable, considering that it is a fully nonlinear scheme.

Normally, the accuracy and convergence of numerical models are tested by using exact (progressive) Stokes waves as a benchmark. For the HOS method, Dommermuth and Yue (1987) studied the maximum absolute error in surface vertical velocity as a function of the order m and the maximum alias-free wave number M for a range of steepness, defined for the Stokes wave as $0.5k(\eta_{\max} - \eta_{\min})$ with fundamental wave number $k = 1$, which is also the definition adopted in the following parts. As reported, for the steepness $\varepsilon \leq 0.35$, the error decreases exponentially fast with both increased m and M. However, if $\varepsilon = 0.40$, which is approximately 90% of the Stokes limiting steepness, the expected convergence rate with respect to m cannot be realized.

On the contrary, the Stokes wave in Fig. 7.2, obtained based on conformal mapping with a precise calculation algorithm (Chalikov and Sheinin 1998; Chalikov 2007), has been developed in CS model, which can achieve an accuracy of 10^{-11} even up to the steepness $\varepsilon = 0.42$ and propagate for an extended period without any distortion. Slight oscillations of the Fourier amplitudes in the case of $\varepsilon = 0.427$) can be observed, reflecting the approach to the critical value of 0.429, corresponding to the Stokes wave with the maximum energy. For $\varepsilon = 0.43$ and $\varepsilon = 0.44$, the unsteadiness of Stokes waves is detected, manifested in the finite-amplitude oscillations of the Fourier components but not breaking (Chalikov and Sheinin 2005).

Thus, it can be argued that the fully nonlinear CS model also has high accuracy and good convergence. To further study its property, some numerical experiments published in the work of Dommermuth and Yue (1987) will be borrowed directly and compared with the corresponding results of the fully nonlinear CS model.

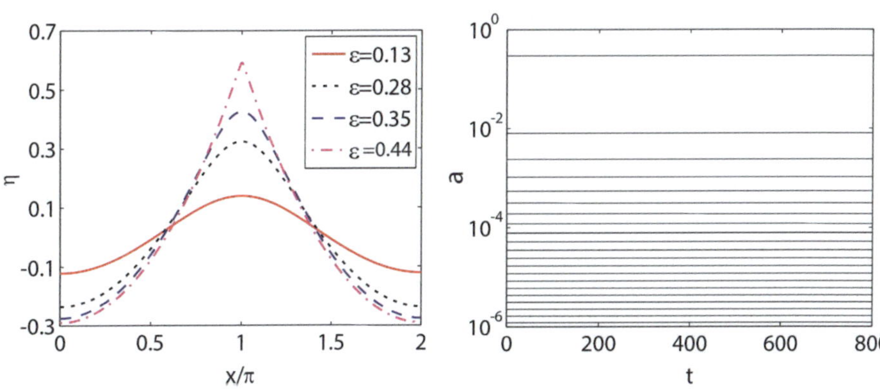

Fig. 7.2 Profiles of Stokes waves (left); long-term evolution of amplitudes of the first 160 constituents of Stokes wave with $\varepsilon = 0.42$ (right, every 8th constituent is shown)

7.2.2 Modulation of Stokes Wave Train Due to Type I Instability

With the HOS method, Dommermuth and Yue (1987) studied the type I instability of a Stokes wave train and compared it with similar numerical results obtained by extended fourth-order Zakharov's equation. For their Airy wave steepness $\varepsilon = 0.13$, the wave numbers of the most unstable class I modes are approximately $\pm 22\%$ of the fundamental.

To compare with the former work, the same Stokes wave is used as the initial condition, i.e., $[\varepsilon, k_0] = [0.13, 9]$, modulated by two Airy sideband waves ($k_1 = 7$ and $k_2 = 11$ are chosen so that integral numbers of the sideband modes can be fitted into the computational domain):

$$\begin{aligned}\eta(x,0) = \eta_0[0.13, 9] + 0.1a_0\cos\left(7x - \pi/4\right) \\ + 0.1a_0\cos\left(11x - \pi/4\right),\end{aligned} \tag{7.27}$$

$$\begin{aligned}\phi^s(x,0) = \phi_0^s[0.13, 9] + \frac{0.1a_0}{\sqrt{7}}\mathrm{e}^{7\eta}\sin\left(7x - \pi/4\right) \\ + \frac{0.1a_0}{\sqrt{11}}\mathrm{e}^{11\eta}\sin\left(11x - \pi/4\right),\end{aligned} \tag{7.28}$$

where a_0 is the amplitude of the fundamental mode. The number of free wave modes in the CS model is $M = 64$, the same as that given in the HOS method where the approximated nonlinearity is up to order $m = 4$, but different from the fourth-order simulation governed by Zakharov's equation, for which only a limited approximation with five free waves is used. Note that the time step is 1/50 of the fundamental wave period T.

To describe the discrepancy clearly, their results related to the time histories of the fundamental ($k_0 = 9$), subharmonic ($k_1 = 7$), and superharmonic ($k_2 = 11$) are replotted and compared with the corresponding result given by the CS model in Fig. 7.3. All numerical models predict a first minimum of the fundamental near $t \approx 60T$, which closely corresponds to the timescale for type I interactions, T/ε^2.

Within a short-term evolution such as $t < 60T$, compared with the fourth-order Zakharov's equation, the CS model agrees much better with the HOS method in the amplitude of each harmonic relative to the initial amplitude of the fundamental, revealing that the higher-order nonlinear terms can lead to a higher increase in the subharmonic mode ($k_1 = 7$) and a further decrease in the fundamental mode.

The simulation, governed by the CS model, neither breaks down after $t/T \approx 140$ as appeared in the HOS method nor repeats the same evolution (Fermi–Pasta–Ulam recurrence) as shown in the fourth-order Zakharov's equation. It presents an approximate recurrence with the peak frequency gradually downshifting to the lower subharmonic ($k_1 = 7$) mode.

Besides, the time for the fundamental mode to re-dominate the evolutionary process decreases with the increased nonlinearity in these models. Thus, it can be concluded

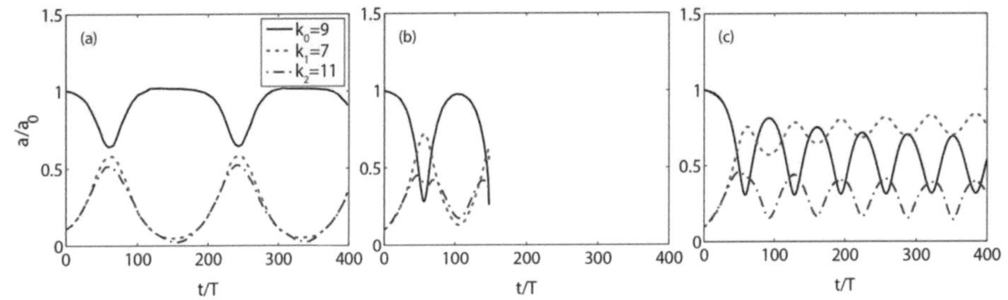

Fig. 7.3 Time histories of the amplitudes of the fundamental ($k_0 = 9$), subharmonic ($k_1 = 7$) and superharmonic ($k_2 = 11$) modes relative to the initial amplitude of the fundamental for an evolving Stokes wave train: **a** from Stiassnie and Shemer (1987); **b** from Dommermuth and Yue (1987); **c** obtained with CS model

that the effect of the higher-order nonlinearity is more significant and pronounced in the long-term evolutionary process.

Figure 7.4 shows the initial surface elevation at time $t/T = 0$ when the fundamental dominates, and the extreme surface elevation appeared in the CS model at time $t/T = 60$ when the sidebands dominate as well as the corresponding one obtained with HOS method at time $t/T = 57$. Obviously, in the CS model, the maximum local wave and its slope are not as large as those that appeared in the HOS method, where the maximum local wave slope reaches almost four times ($\eta_x \approx 0.6$) that of the initial wave.

Moreover, in the CS model, total energy is conserved during the entire evolutionary process. Still, the HOS method presents an energy change to various degrees, particularly at the moment corresponding to the maxima in the sideband perturbations (suddenly decreased by nearly 1% the first time and more than 2% the second time).

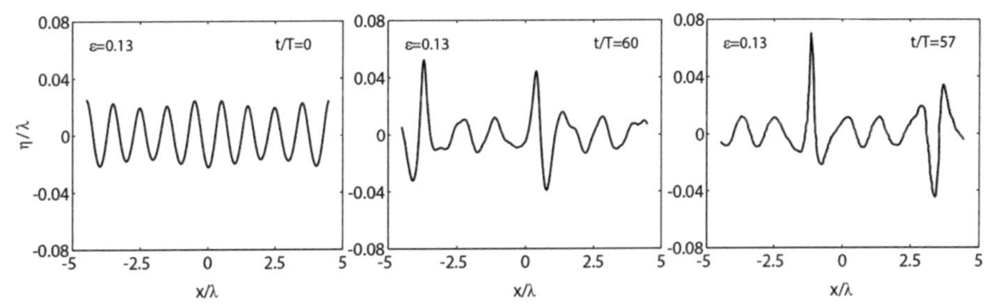

Fig. 7.4 Instantaneous free surface elevations of an evolving Stokes wave train (fundamental period T and wavelength λ): the initial surface elevation at time $t/T = 0$; the moment the sidebands dominate at time $t/T = 60$ in CS model; the moment the sidebands dominate at time $t/T = 57$ in HOS method, borrowed directly from Dommermuth and Yue (1987)

To further explore this kind of evolution, the same numerical experiments with different initial fundamental steepness and a different number of free wave modes are also given in Fig. 7.5. Within the same number of free wave modes M, i.e., for each row in Fig. 7.5, the downshifting tendency of peak frequency and the exchange rate of wave energy among free modes will substantially decrease with the decreased initial wave steepness while the recurrence time is dramatically enlarged with more significant periodicity.

For each column with the same initial fundamental steepness, the influence of the number of free wave modes can be ignored before the first time the sidebands dominate. If the nonlinearity is not strong, such as $\varepsilon = 0.10$ in the third column of Fig. 7.5), no significant discrepancy can also be observed in the following evolutionary stage.

However, as the nonlinearity increased, e.g., $\varepsilon = 0.11$ in the second column of Fig. 7.5, more free wave modes led to a more extended recurrence period and less difference between the evolutions of the two sideband waves. Suppose the nonlinearity is large enough, as shown in the first column. In that case, the increase in the number of free waves will still but not significantly contribute to the more extended recurrence period.

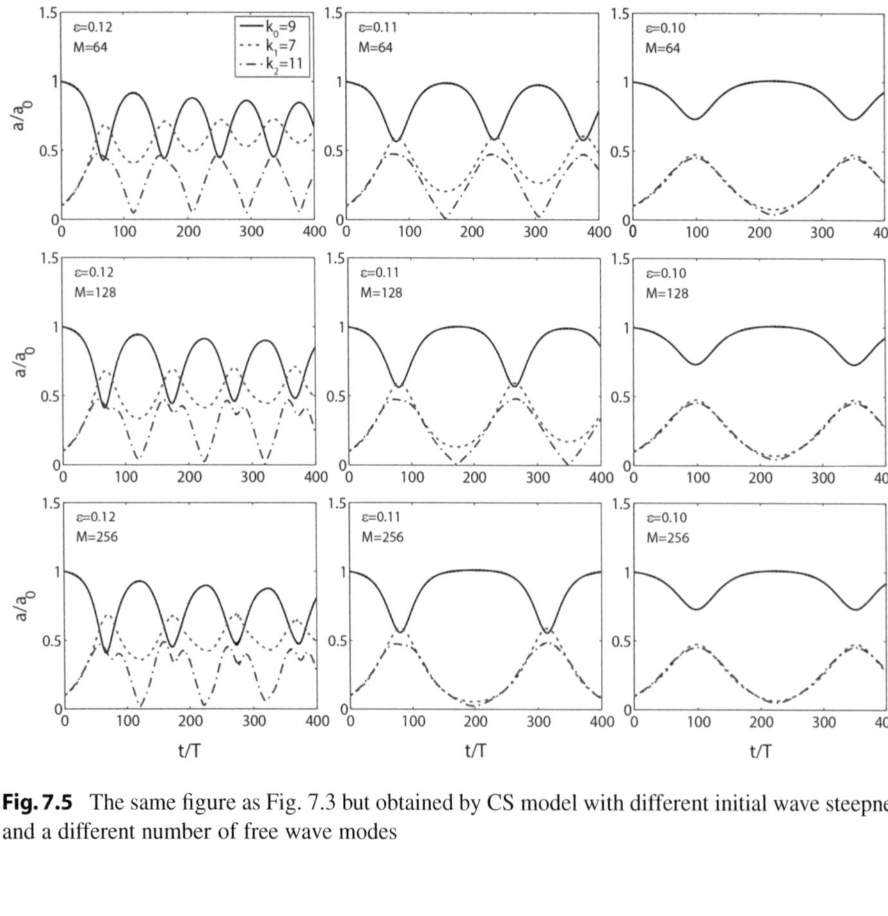

Fig. 7.5 The same figure as Fig. 7.3 but obtained by CS model with different initial wave steepness and a different number of free wave modes

Generally speaking, the evolutionary tendency of the superharmonic mode ($k_2 = 11$) agrees with that of the subharmonic mode ($k_1 = 7$). However, another small variation during each peak stage is detected in the case of $M = 128$ and 256, consistent with the fundamental's evolutionary tendency.

In Fig. 7.6, the instantaneous surface profiles of the evolving Stokes wave train are displayed when the sidebands dominate for the first time. As the initial wave steepness decreases, the Stokes wave train is modified into two approximately symmetric wave packets, especially in the case of weak nonlinearity, e.g., $\varepsilon = 0.10$, of which the carrier wave numbers are the superharmonic and subharmonic modes, respectively.

As indicated before in the time histories of the amplitudes in Fig. 7.5, the number of free wave modes does not have a significant impact at this moment except for the case with solid nonlinearity shown in the first column of Fig. 7.6 where higher crest and deeper trough are observed in the presence of more free wave modes.

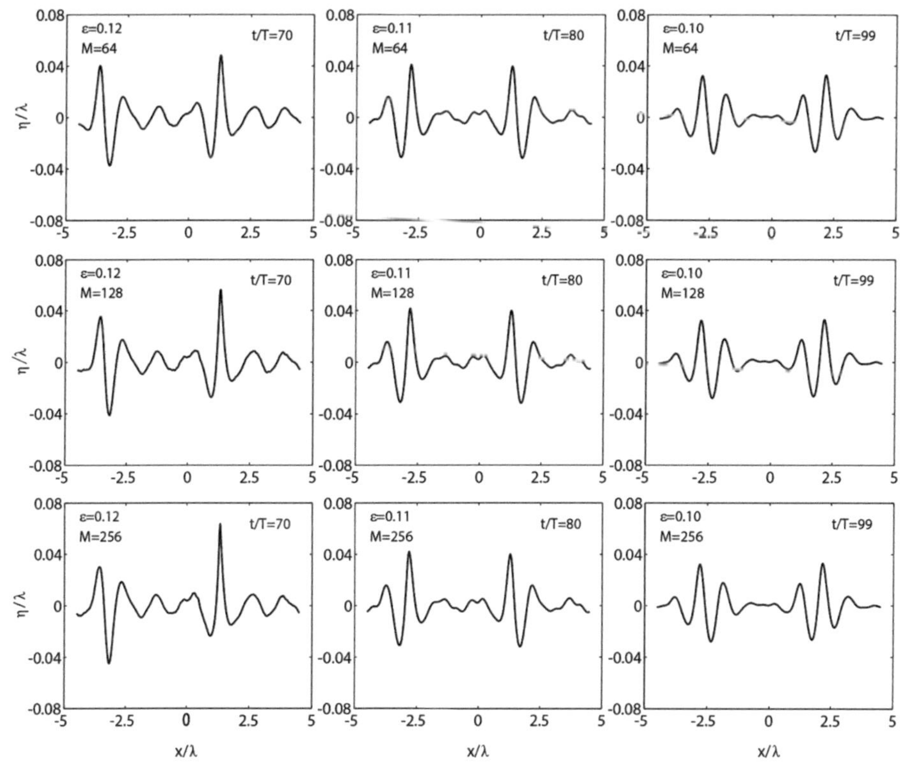

Fig. 7.6 Instantaneous free surface elevations of the evolving Stokes wave train with different initial fundamental steepness and free wave modes at the moment the sidebands dominate

7.2.3 Evolution of Wave Packet

In this section, the fully nonlinear CS model is compared with the HOS method in simulating one of Su's experiments, which initially has a steepness of $\varepsilon = 0.15$ and a packet containing approximately five fundamental waves at the beginning (Su 1982).

To compare with the laboratory experiment, a Stokes wave train with 15 waves in the computational domain, i.e. $[\varepsilon, k] = [0.15, 15]$ is used and modified with a tapering function of the flowing form:

$$F(x; \kappa, x_b, x_e) = 0.5\{\tanh[\kappa(x - x_b)] - \tanh[\kappa(x - x_e)]\}, \tag{7.29}$$

where the parameter κ measures the steepness of the taper at the beginning and end positions, x_b and $x_{e,}$ respectively, of the resulting envelope.

To compare with the HOS method exactly, the same initial condition that approximately gives the desired five waves in the packet has been adopted.

$$\eta(x, 0) = F\left(x; \frac{30}{\pi}, -\frac{\pi}{3}, \frac{\pi}{3}\right)\eta_0[0.15, 15], \tag{7.30}$$

$$\phi^s(x, 0) = F\left(x; \frac{30}{\pi}, -\frac{\pi}{3}, \frac{\pi}{3}\right)\phi_0^s[0.15, 15]. \tag{7.31}$$

It needs to be mentioned here that since the group velocity is roughly half of the phase velocity in deep water, a wave gauge would measure ten waves in Eq. (7.30). The parameters in CS model are the same as those used in HOS method, i.e., $M = 256$ and $T/\Delta t = 40$.

Note that the HOS method to include the higher-order nonlinearity m is set up to 6 in the numerical simulation. The ideal filter is also applied to allow the computations to continue after the waves become locally too steep.

The simulated results in the CS model are presented in the right column of Fig. 7.7, compared with those computed by the HOS method and Su's wave probe measurements, which are just sketch maps because of no vertical scale in the original work of Su (1982).

Overall, the two kinds of numerical simulations are consistent with each other and able to catch the fundamental characteristics of the evolution of a wave packet. The agreement with experiments appears to improve as the wave group travels downstream in the wave tank. Both numerical simulations confirm that the wave envelope becomes forward-leaning after a short distance. Then, the wave group splits into two packets with the smaller group trailing, which is qualitatively in conformity with the laboratory observation.

However, it has to be admitted that a certain degree of difference can be observed between these two numerical results, especially in the aspect of total energy in the wave system. As indicated in Fig. 7.8, almost 20% of the power has been removed from the

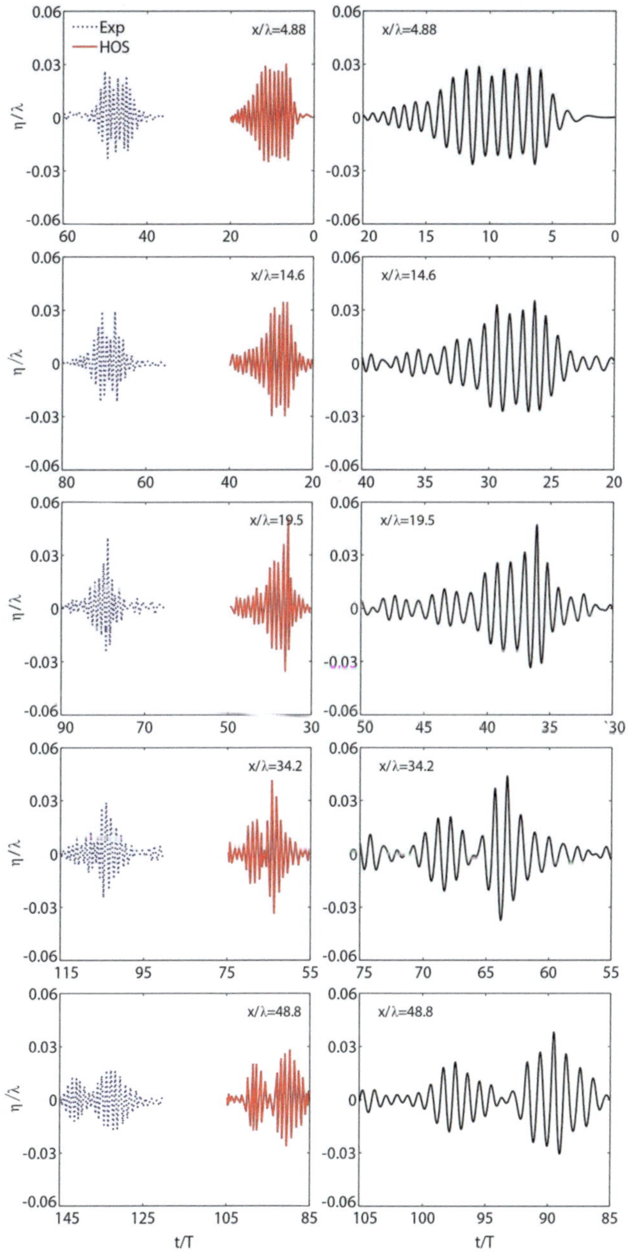

Fig. 7.7 Comparison of simulations of CS model (right column) and HOS method with laboratory experiments (left column, from Dommermuth and Yue 1987) for the free surface elevation of an evolving wave packet (fundamental period T and wavelength λ) at positions $x/\lambda = 4.88$, 14.6, 19.5, 34.2 and 48.4, respectively. Note that the left column is roughly reproduced due to the low resolution in the published paper

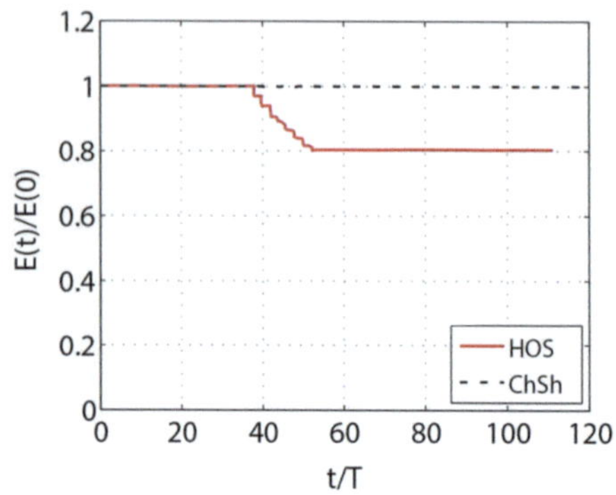

Fig. 7.8 Comparison between CS model and HOS method for the computed total energy of the evolving wave packet; the red solid line is from Dommermuth and Yue (1987)

system in the evolutionary process controlled by the HOS method due to the use of an ideal filter, which is applied whenever the total energy changes by more than 1%.

7.2.4 Collision of Two Wave Groups

To demonstrate the performance of CS model for a larger-scale problem, the head-on collisions are considered between two wave groups, for which the wavelength of the waves in one group is twice that of the other.

For each wave group, the initial steepness of the waves is $\varepsilon = 0.16$. The same initial parameters are used to compare with the HOS method, $M = 512$ and $T_S/\Delta t = 40$, where T_S is the fundamental period of the shorter waves.

The initial conditions are given as:

$$\eta(x, 0) = F\left(x; \frac{25}{\pi}, -\frac{9\pi}{25}, -\frac{3\pi}{25}\right)\eta_0[0.16, 50] + F\left(x; \frac{25}{2\pi}, \frac{6\pi}{25}, \frac{18\pi}{25}\right)\eta_0[0.16, 25], \tag{7.32}$$

$$\phi^s(x, 0) = F\left(x; \frac{25}{\pi}, -\frac{9\pi}{25}, -\frac{3\pi}{25}\right)\phi_0^s[0.16, 50] - F\left(x; \frac{25}{2\pi}, \frac{6\pi}{25}, \frac{18\pi}{25}\right)\phi_0^s[0.16, 25]. \tag{7.33}$$

As shown in Fig. 7.9a, the initial profiles have approximately six waves in each group and propagate to opposite directions. After the interaction, the new free surface profiles are presented in Fig. 7.9b at $t = 31T_S$ when the groups have emerged on opposite sides.

To evaluate the effect of the inter-group interactions, each simulation with only a single propagating group is repeated and plotted in Fig. 7.9b, being marked by the red and blue dash lines for the same moment. In agreement with the observation in the HOS method,

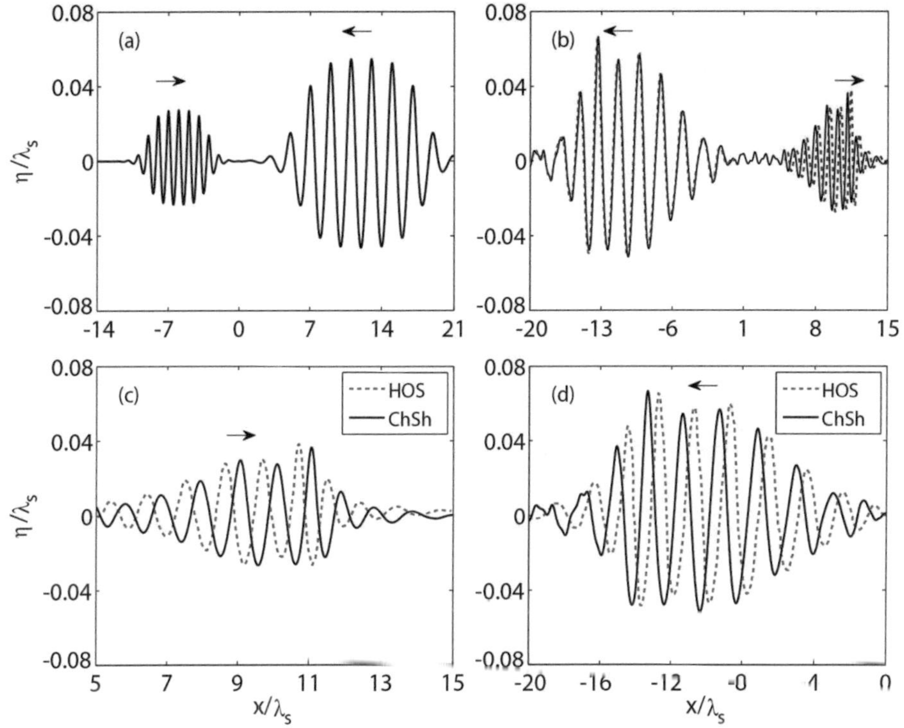

Fig. 7.9 Free surface elevations of two colliding wave packets of different carrier periods T_L and T_S and wavelengths λ_L and λ_S with $\lambda_L = 2\lambda_S$. **a** Initial profiles before the collision; **b** final surface profiles ($t/T_S = 31$) of the short-wave group and long-wave group, compared to those which have evolved independently (blue and red dash lines); **c** comparison between CS model and HOS method for the final surface profiles ($t/T_S = 31$) of the short-wave after group collision; **d** the same as **c** but for the long-wave group

the interaction between the groups is also relatively small in a fully nonlinear CS model, even though the effects of nonlinearity are appreciable as reflected by the steepened fronts and the characteristic wedge-like shapes of the individual envelopes.

It is confirmed once again that, even for the fully nonlinear interactions, the only effect of the collision is just a small shift in the position of the envelope due to a decrease in the group speed. The effect is more pronounced for the short-wave group, supporting the prediction of Longuet-Higgins and Phillips for two third-order Stokes wave trains travelling in opposite directions and generalizing the observations of Oikawa and Yajima for interacting solitons with different carrier frequencies in the fully nonlinear model.

To see the difference between the sixth-order nonlinear ($m = 6$ in HOS method) and fully nonlinear (in CS model) interactions, the corresponding short-wave and long-wave groups after collision have been compared in Figs. 7.9c, d, respectively. Since the evolutionary time is relatively short, i.e., $31T_S$, there is no significant discrepancy in the free surface profiles after the collision, except that both wave groups' speeds are higher in the CS model than in the HOS method.

According to the above comparisons in this section, it can be further inferred that the CS model and the HOS method will not make a significant difference, even in the case of simulating the more realistic sea state characterized by the JONSWAP power spectrum.

This is also confirmed in our numerical tests. The corresponding results are not presented in the former study associated with the HOS method in Chap. 6 to avoid the problems of superfluous repetition of published works and overlapping with the following discussion in Sect. 7.3.

7.3 Comparison Between Dysthe Equation and CS Model

The laboratory experiments of unidirectional waves that will be compared were also performed at the Marintek wave basin in Trondheim, Norway (Zhang et al. 2016a). Descriptions of the facilities and experiments can be found in the former Sect. 5.2 (see Table 2.11).

In numerical simulation, the CS model can only provide temporal evolution of an initial wave field imposed at $t = 0$, and thus, problems will arise when comparing numerical results with laboratory experiments measured by wave gauges. In the following analysis, the method used in Chap. 5 is still adopted here. The numerical simulation of 2D temporal MNLS equation, also named Dysthe equation, will be presented in this section.

7.3.1 Statistical Parameters

The spatial variations of skewness and kurtosis coefficients are plotted in Fig. 7.10 and Fig. 7.11, respectively. Normally, in the case of wave series with strong nonlinearity, higher, more peaked crests and shallower, more rounded troughs can be observed. As discussed before in Sect. 5.4, the asymmetry of the wave profile is induced by two different types of nonlinear effects: the major one dominated by bound waves, the other weak one contributed from free waves (Onorato et al. 2005; Mori et al. 2007; Toffoli et al. 2008a).

In Fig. 7.10, both numerical simulations can capture the variation of asymmetry, but a significant discrepancy can be detected in the locations larger than $20L$ or $40T$, where the variability is very large in the experiments indicated by the longer error bar. Nevertheless, both models are consistent with each other and capable of catching the fundamental characteristics of skewness.

For the coefficient of kurtosis, if BFI is small, e.g., Fig. 7.11a, the CS model shows a better agreement with the laboratory observation than the MNLS equation. However, if the

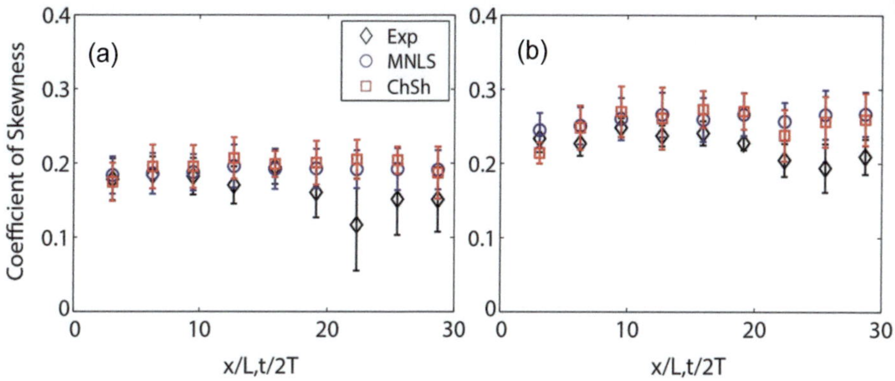

Fig. 7.10 Spatial variation of coefficient of skewness

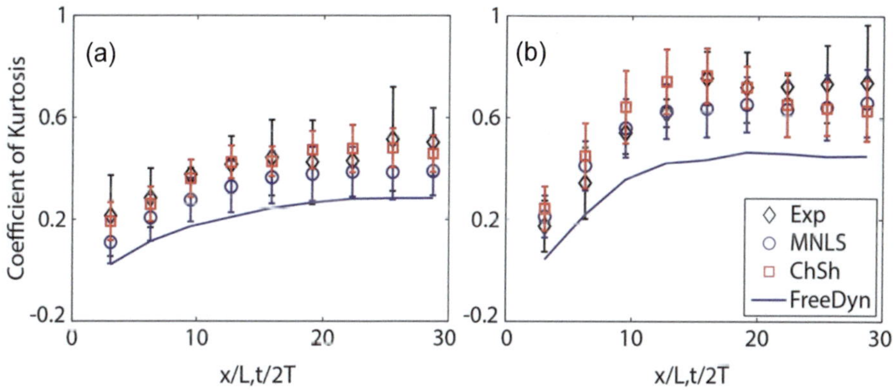

Fig. 7.11 Spatial variation of coefficient of kurtosis. The blue solid line corresponds to the contribution of the dynamic part derived from the MNLS equation

nonlinearity increases, such as in Fig. 7.11b, both numerical models still work reasonably well to catch the evolutionary tendency but present different levels of discrepancy from the experiments in the wave basin.

The contribution to the coefficient of kurtosis can be divided into the dynamical and bound harmonics parts separately, as shown in the work of Annenkov and Shrira (2009), where the temporal evolutions of each part and their sum are simulated with the kinetic equation. The numerical results of Annenkov and Shrira can be quantitatively compared with the laboratory experiments performed in a large wave tank for different types of wave spectra within a range of spectral widths (Shemer and Sergeeva 2009; Shemer et al. 2010a). In their experiments, the contributions to kurtosis were extracted and compared with the theoretical predictions, consistent with the results shown in Fig. 7.11. Moreover, their laboratory observation reveals that the spectral width plays an important role in the evolution of the coefficient of kurtosis.

Concerning the relationship between coefficients of skewness and kurtosis in Fig. 7.12, different experimental data have been analyzed herein, but similar conclusions can be derived as those in Sect. 4.3. Therefore, it can be argued that there is no clear relationship between these two coefficients in the presence of third-order nonlinearity. As to the numerical simulation, the relationships revealed by the CS model and the MNLS equation remain consistent with the experimental results. Until now, the CS model does not present many advantages compared to the MNLS equation.

In Fig. 7.13, the fourth-order normalized cumulant Λ and Λ_{app} still follows the principle and rule derived in Sect. 4.3. As pointed out before, both experiments and simulations contradict the second-order theoretical conclusion, which states that $\Lambda > \Lambda_{app}$ if $\lambda_{40} < 3\lambda_{22} < \lambda_{04}$ (Tayfun and Fedele 2007). According to the previous work (Zhang et al. 2014), it has been inferred that the nonlinear modulation does not contribute too much to the discrepancy between these two parameters in the absence of bound wave effects.

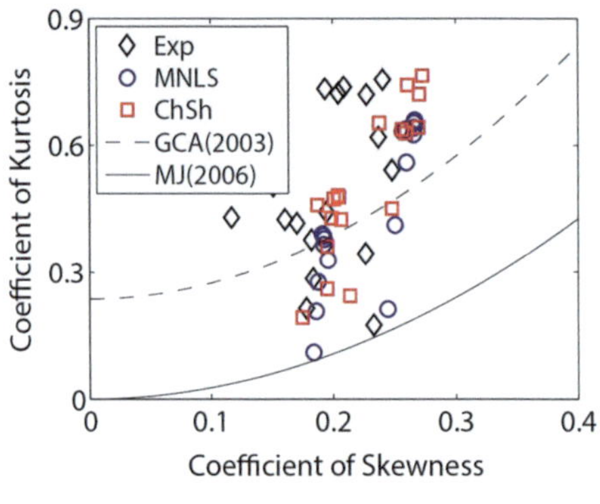

Fig. 7.12 Relationship between coefficients of kurtosis and skewness. The dashed line is obtained by Guedes Soares et al. (2003), while the solid line is proposed by Mori and Janssen (2006)

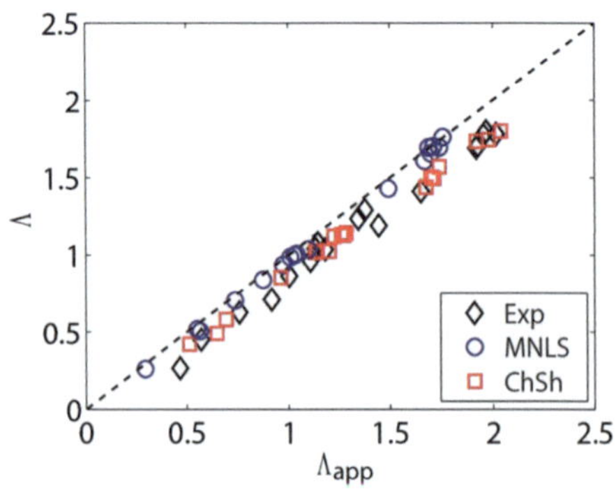

Fig. 7.13 Relationship between Λ and Λ_{app}

In other words, the difference between these two parameters is actually induced by the combination of nonlinear modulation and bound wave effects. Moreover, this conclusion is further exploited here by the difference in the numerical simulations: the MNLS equation can only identify a little difference, while the CS model, in contrast, provides an exact description of their relationship. Now it can be said that the nonlinear modulation greatly enlarges the higher-order bound wave effects, which consequently leads to $\Lambda < \Lambda_{app}$.

7.3.2 Distributions Related to Free Surface Elevation

Figure 7.14 represents the probability density functions of the surface elevations measured at Gauge 6 for case A and at Gauge 5 for case B, where the nonlinear modulation has already been fully developed in both random sea states.

The former chapters have discussed analysis associated with the involved nonlinear dynamics and will not be repeated here. Obviously, the MNLS equation is not fully nonlinear, but it can still grasp the surface elevation statistics reasonably well as performed by the fully nonlinear CS model.

Furthermore, it can be argued that a higher-order nonlinear theoretical model must be proposed to better fit the tails of wave surface distribution. However, it should also be mentioned here that the directional spreading in the 3D case can significantly decrease the effect of nonlinear modulation (Stansberg 1994; Waseda 2005); thus, the deviation between experiment and second-order theoretical prediction may not be as large as above.

Since the corresponding spatial variations of the exceedance distributions are almost the same as those presented in the other laboratory experiments (Zhang et al. 2014, 2015), the study in the following part will only focus on the same single location as that in the

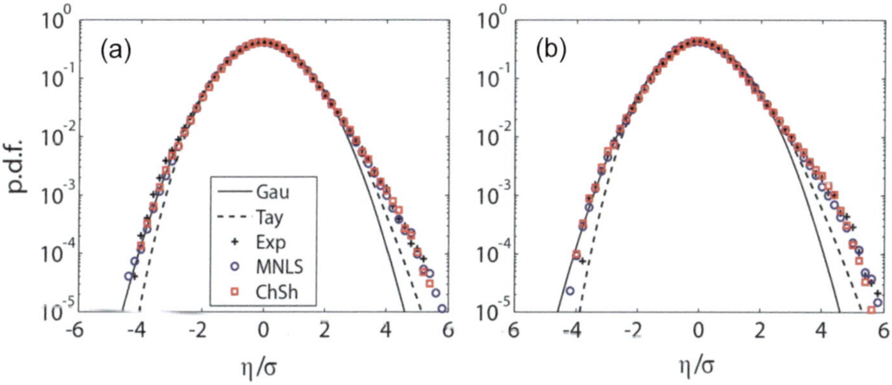

Fig. 7.14 Probability density function of the surface elevations in logarithmic scale located at $x/L = 19.1$ for case **a** and at $x/L = 15.9$ for case **b**

analysis of surface elevation but for wave crest, trough and height now. Besides, the model predictions from linear to nonlinear are also plotted in Figs. 7.15 and 7.16 for theoretical analysis of the experimental data.

To be convenient in comparison of numerical simulations with laboratory observations, the exceedance probability is divided into three parts in purpose: low probability (10^{-4}, 10^{-3}), intermediate probability (10^{-3}, 10^{-1}) and high probability (10^{-1}, 10^{0}).

Apparently, in Fig. 7.15, only the third-order GC model can reasonably estimate the wave crest distribution for both random sea states. Compared with the second-order model (NB) prediction, it reveals that the nonlinear modulation mainly contributes to the formation of larger crest amplitudes. Concerning the numerical simulations, the fully nonlinear CS model gives a good simulation of laboratory observations at all probability levels,

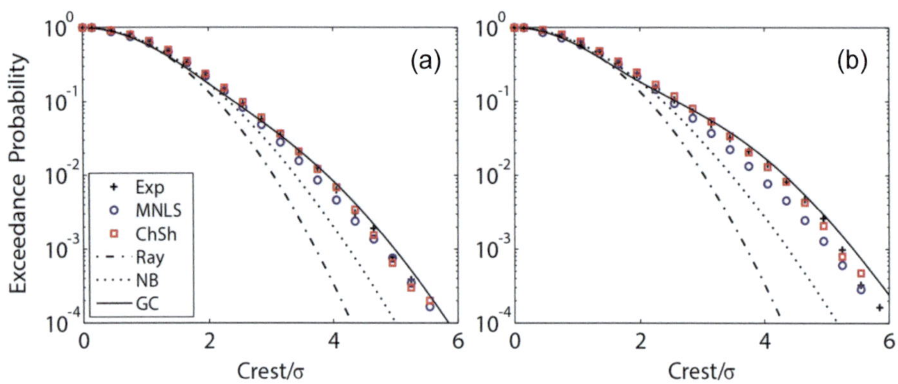

Fig. 7.15 Exceedance distribution of wave crests located at $x/L = 19.1$ for case **a** and at $x/L = 15.9$ for case **b**

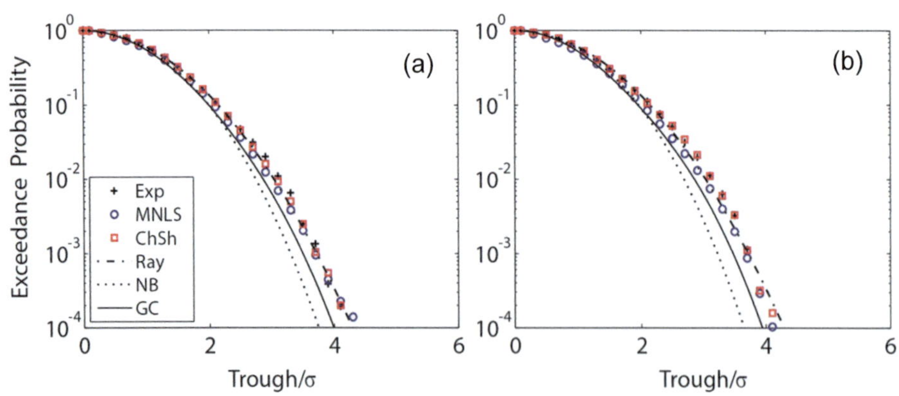

Fig. 7.16 Exceedance distribution of wave trough located at $x/L = 19.1$ for case **a** and at $x/L = 15.9$ for case **b**

while the MNLS equation underestimates the experimental results at the intermediated probability, i.e. (10^{-3}, 10^{-1}), especially pronounced in the presence of a high degree of nonlinearity such as case B in Fig. 7.15.

In second-order wave theory, it is relatively well established that the troughs of nonlinear wave profiles are flatter than those in a Gaussian random process. Therefore, their statistical distribution is expected to deviate from the linear Rayleigh model. However, as presented in Fig. 7.16, the trough distributions of mechanically generated unidirectional waves are fitted very well by the linear Rayleigh function, which agrees with the numerical simulations of the CS model and the MNLS equation.

As a result, it can be inferred that the interesting linear Rayleigh distribution of wave trough is the result of the opposite actions of bound wave effects (shallower) and nonlinear modulation (deeper). Moreover, similarly to what has been observed in the exceedance distribution of wave crest, the CS model behaves a little better than the MNLS equation in the intermediate probability for wave trough exceedance distribution again.

What needs to be pointed out is that the comparison only represents long-crested wave trains. Field measurement has shown that the Rayleigh distribution overestimates the trough distribution, and the second-order theory can describe the crest distribution (e.g., NB). Thus, the deviation will not be very noticeable in a real ocean environment, as presented in Figs. 7.15 and 7.16.

Since the difference between numerical simulations is mainly reflected in the range from 10^{-3} to 10^{-1}, being more evident in the case with higher degree of nonlinearity, it can be further inferred that the MNLS equation can catch the third-order nonlinear effect, indicated at the low probability level, i.e. (10^{-4}, 10^{-3}), while the more subtle features that involve not only the nonlinear modulation but also the higher-order bound wave effects can only be captured by the fully nonlinear CS model as demonstrated at the intermediate probability level, i.e. (10^{-3}, 10^{-1}).

Figure 7.17 presents the exceedance distribution of wave heights, defined as the sum of trough depression and crest elevation in an individual wave. It is well known that second-order nonlinearity does not significantly affect the wave height distribution. Moreover, the fact that the observed statistics in the experiments are only fitted perfectly by the third-order nonlinear GC model of Eq. (3.5), further confirms that the generation of large amplitude waves is mainly attributed to the quasi-resonant four-wave interaction rather than due to phase-locked modes (Stokes contribution).

Finally, it is distinct that a significant discrepancy can be observed between numerical simulations and experimental results, starting from the intermediate probability and becoming pronounced to the low probability. Part of the difference is due to the indirect comparison between simulations of temporal evolution models and gauge measurements for spatial evolution, which is supported by the fact that the numerical results, which are slightly underestimated by the Rayleigh distribution function to various degrees depending on the nonlinearity of the wave series, are in agreement with those performed by the HOS method such as in the work of Toffoli et al. (2008a).

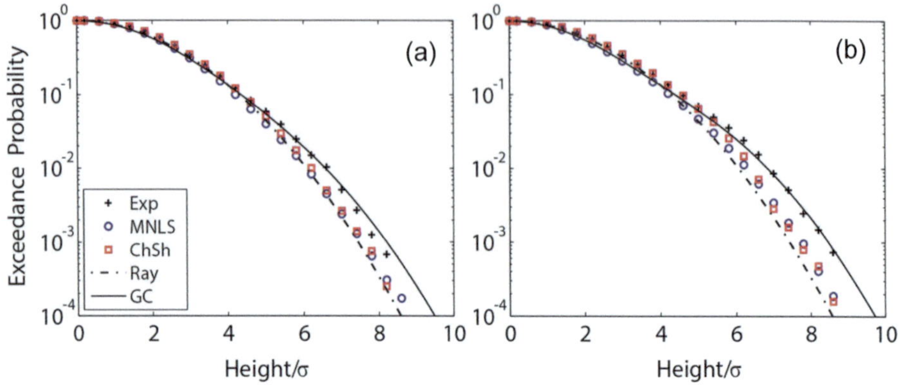

Fig. 7.17 Exceedance distribution of wave heights located at $x/L = 19.1$ for case **a** and at $x/L = 15.9$ for case **b**

7.4 Conclusion

It has been established that the CS model can simulate breaking conditions when the surface becomes a multi-valued function of the horizontal coordinate. Formally, the simulation continues up to the moment when an overturning volume of water touches the surface. Besides, in simulating the progressive Stokes wave, it can achieve an accuracy of 10^{-11} even up to the steepness $\varepsilon = 0.42$ and propagate for a long period without any distortion, while the simulation governed by the HOS method is reported to fail if $\varepsilon > 0.35$.

In modelling the modulation of a Stokes wave train with initial steepness $\varepsilon = 0.13$, all numerical models can predict a first minimum of the fundamental around $t \approx 60T$, which closely corresponds to the timescale for type I interactions T / ε^2. However, the simulation in the CS model neither breaks down after $t/T \approx 140$ as displayed in the HOS method nor repeats the same evolution (Fermi–Pasta–Ulam recurrence) as revealed in the fourth-order Zakharov's equation. It behaves in an approximate recurrence, with the peak frequency gradually downshifting to the lower subharmonic mode. Moreover, the maximal local wave and its slope in the CS model are smaller than that given in the HOS method when the sidebands first dominate the evolutionary process. Compared to the effect of initial fundamental steepness, it seems that for the fully nonlinear CS model, the influence of the number of free wave modes can be ignored in the short-term evolution.

Overall, both the CS model and HOS method are qualitatively consistent with Su's laboratory observation and capable of catching the primary characteristics of the evolution of a wave packet. Furthermore, during the whole process, the total energy was conserved in the CS model, while almost 20% of the initial energy was removed from the system controlled by the HOS method due to a filter applied to the simulation.

Concerning the nonlinear interaction between two wave groups, the fully nonlinear CS model indicates a similar phenomenon as that illustrated in the HOS method, that is, the only effect of the collision is just a small shift in the position of the short-wave envelope due to a decrease of wave group speed. As a result of a short-term evolution, the discrepancy in the free surface profile is not conspicuous, but the higher group speeds in the CS model can be detected.

Regarding the more realistic sea state, the CS model attempted to simulate the 2D laboratory experiments in Marintek. Compared with the simulation of the MNLS equation's temporal version, the fully nonlinear CS model can present a little better performance, particularly regarding the fourth-order normalized cumulants and the intermediate probability of wave crest and wave trough distributions. Except for the exceedance distribution of wave height, the critical features of mechanically generated long-crested waves can be captured by the CS model and to some degree, by the MNLS equation.

Regarding the coefficients of skewness and kurtosis, various discrepancies are encountered between numerical simulations and experimental results. However, both numerical models have successfully reproduced the fundamental characteristics of asymmetry of wave profiles and the formation of larger amplitude waves. Besides, Λ becomes smaller than Λ_{app} as a result of the combination of nonlinear modulation and bound wave effects, clearly elucidated in the numerical simulation governed by the fully nonlinear CS model rather than by the MNLS equation.

Since the wave crest distribution is only fitted reasonably well by the third-order GC model, it means that the nonlinear effect among free wave modes mainly contributes to the formation of larger crest amplitude, which can be exactly captured by the fully nonlinear CS model at the low probability level, that is, the range from 10^{-3} to 10^{-1}. In addition, the fully nonlinear CS model does present a certain degree of advantage the MNLS equation in simulating wave trough, especially in the case with strong nonlinearity.

Even though the deviations of wave crest and wave trough from second-order wave theory can be explored by the numerical simulations performed with the CS model and the MNLS equation, the discrepancies between experiments and simulations are evident in wave height distributions, where the Rayleigh function slightly underestimates the simulated larger wave heights to various degrees but are in line with the work of former researchers who applied the HOS method in simulating similar nonlinear wave series. Such discrepancy can be partially attributed to the indirect comparison between numerical simulations and laboratory observations.

8 Effect of Initial Conditions on Extreme Wave Formation

8.1 Disturbance of Noise in Deterministic Wave Train

According to the weakly nonlinear NLS theory, the evolutionary process of a breather wave depends on the product of two physical parameters, i.e. the wave steepness and the number of waves in one modulation (see, e.g., Babanin et al. 2019; Slunyaev and Shrira 2013, and reference therein). In this subsection, both parameters are selected to guarantee the maximum wave amplitude attainable is equal to three in theory, and the same Peregrine breather solution will evolve in both regular and irregular background waves.

8.1.1 Regular Background Waves

In Fig. 8.1, one case demonstrates that the wave crest height can reach three times the initial carrier wave amplitude even within the framework of fully nonlinear potential equations. The initial surface elevation is essentially sinusoidal, depicted in Fig. 8.1a with the amplitude of the carrier wave $a = 0.75$ cm and the peak frequency $\omega_0 = 10.68$ rad/s, leading to the steepness $ak_0 = 0.087$ under deepwater condition. These values are set to continue the study of the corresponding laboratory experiment (Chabchoub 2016), whose initial setting is to allow for sufficient distance to develop such evolution while ignoring the effect of surface tension.

As shown in Fig. 8.1b, a significant increase in the amplitude can be observed in the middle part of this periodic wave, which will disappear and approach the amplitude of the background wave if far away enough from the location of nonlinear focusing. According to the analytical Peregrine breather solution, the amplitude of the carrier wave can be amplified as large as three times the initial one (red dotted line). The fully nonlinear

H. Zhang and C. Guedes Soares, *Numerical Modelling of Extreme Waves*, Synthesis Lectures on Ocean Systems Engineering, https://doi.org/10.1007/978-3-031-77084-5_8

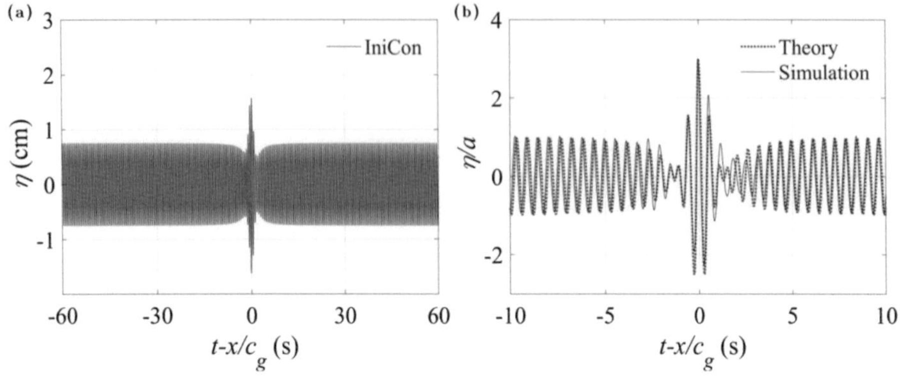

Fig. 8.1 **a** Initial surface elevation of the Peregrine breather solution, 6 m away from the theoretical focusing point; **b** the first crest amplification of three times the carrier wave amplitude, compared with the theoretical prediction

numerical simulation (solid black line) also detects the same amplification factor. Still, the symmetric profile in NLS theory cannot be reproduced any more, in line with the corresponding laboratory observation (Chabchoub et al. 2011).

To further explore the difference between the NLS and fully nonlinear breathers, their whole evolutionary processes are presented separately. The propagation speed of the fully nonlinear breather is underestimated by the linear wave group velocity c_g, revealed by the forward movement of the middle unstable packet in Fig. 8.2b. It seems that this group speed is gradually increased during the nonlinear focusing process ($x < 9.1$ m), and after that, it slowly returns to the linear one once again. For both evolutionary processes, it is clearly seen that the localized wave group contracts considerably and sucks energy sequentially from the back and front waves. The dramatic difference is that for the fully nonlinear breather, most of the accumulated energy is contributed by the back waves with the corresponding energy transfer process irreversible in part.

To demonstrate the whole amplification process of the middle unstable breather packet, the spatial variations of its maximal wave crest amplitude, trough amplitude, crest-to-trough height and envelope amplitude are depicted in Fig. 8.3, respectively. As stated in the weakly nonlinear wave theory (see Fig. 8.3a), the maximum height of the wave envelope and wave crest can reach three times the carrier wave amplitude at the prescribed location, i.e. $x = 6$ m. Moreover, it is indicated that the crest and trough amplitudes and the crest-to-trough height are not monotonically increased as observed for the envelope amplitude but are periodically modulated during evolution.

Compared to the NLS theory, the periodic modulation can also be observed in Fig. 8.3b, but the focus distance increases significantly in the fully nonlinear numerical simulation. Taking the wave crest amplitude as an example, it needs a much longer evolutionary distance ($x = 9.1$ m) to attain the global peak value (marked by an asterisk

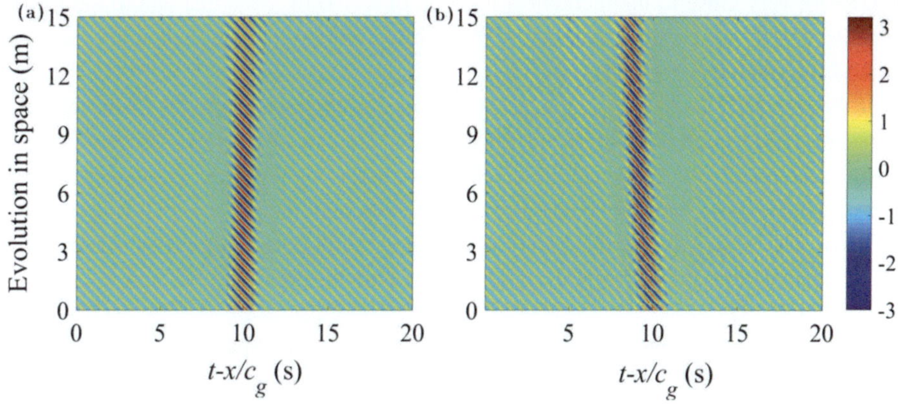

Fig. 8.2 Evolution of the Peregrine breather solution, space–time representation for the surface elevation scaled by a. **a** Theoretical prediction from NLS equation; **b** numerical simulation in the fully nonlinear equation

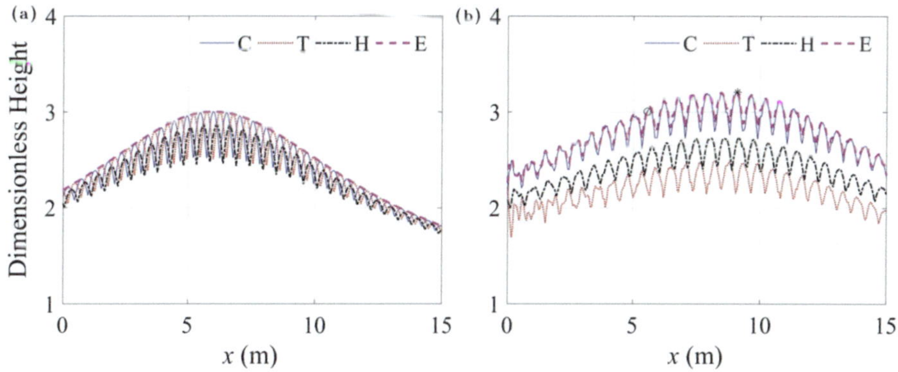

Fig. 8.3 Spatial evolutions of the scaled maximal wave crest amplitude (C), trough amplitude (T), crest-to-trough height (H) and envelope amplitude (E) within the middle unstable breather packet. **a** Theoretical prediction from NLS equation; **b** numerical simulation in the fully nonlinear equation with circle denoting scaled crest first attaining three and asterisk marking its peak value. Crest-to-trough height is scaled by $2a$ while the other variables are scaled by a

in Fig. 8.3b). The noticeable delay of the peak amplitude amplification mainly arises from the elongated (about 11%) individual modulation period. Moreover, the peak amplification index (or the maximum dimensionless height) is larger than the theoretical value of 3 for both wave crest and wave envelop, while it is smaller than the theoretical prediction for wave trough and wave height.

Interestingly enough, the amplification factor of wave crest amplitude attains 3 for the first time at the location of $x = 5.6$ m (see circle in Fig. 8.3b and its detailed profile in

Fig. 8.1b), very close to the theoretical location of $x = 6$ m. Note that due to the insufficient growth in wave trough amplitude, the envelop amplitude is not smoothly increased in the fully nonlinear focusing process.

In both weakly and fully nonlinear cases, the evolutionary processes of the maximal wave crest and trough amplitudes are out of phase. The law of conservation could explain this, i.e. the energy is exchanged iteratively between the highest crest and deepest trough. Consequently, it can be inferred and also confirmed herein that in each modulation period, the local peak values of the crest amplitude, trough amplitude and crest-to-trough height are attained at close but different locations, in line with the previous observation (Slunyaev and Shrira 2013).

For completeness, the extreme value of wave crest amplitude obtained in the fully nonlinear simulation is given in Fig. 8.4a as a function of different initial wave steepness, which is realized by varying the carrier wave amplitude without changing its frequency. To avoid the influence of wave breaking, the upper boundary of initial wave steepness is set to be 0.145. Each case's dimensional extreme crest height is proportional to the initial wave steepness. Meanwhile, the nondimensional one is more complex, i.e., being increased first and then decreased, and is a little unstable around the peak value. It can be further concluded that the maximum amplification factor of the surface elevation attained by the nonbreaking Peregrine breather wave is ~ 3.3, a little bit smaller than 3.4 given by Tanaka (1990), who examined the biggest wave the modulational instability can produce for a given uniform wave train.

Regarding the spatial distance required to develop an extreme wave crest, no clear trend can be observed in Fig. 8.4b. This is mainly caused by the nonlinear effect, which can influence the matching process of the initial conditions and the evolutionary process

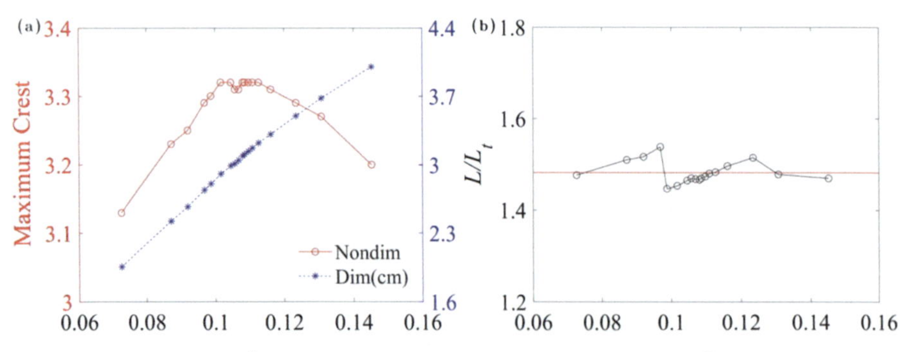

Fig. 8.4 Extreme value of the wave crest amplitude (**a**) and the corresponding evolutionary distance (**b**) derived in the fully nonlinear simulation as a function of the initial wave steepness. The left and right ordinates in **a** correspond to the nondimensional (scaled by the initial carrier wave amplitude) and dimensional crest height, respectively. The evolutionary distance in **b** is scaled by the theoretical value $L_t = 6$ m

of breather dynamics. Thus, the growing distance is becoming more sophisticated with the increased nonlinearity in the numerical simulation (Zhang et al. 2021). Anyway, the required distance is always larger than the theoretical one, with a mean value of $1.48L_t$ marked by the red line in Fig. 8.4b.

Thus, it can be inferred that all discrepancies between theoretical predictions and numerical experiments mainly stem from the higher-order linear and nonlinear effects, ignored in the simplified nonlinear Schrödinger equation. The conjecture can also be partly confirmed by the work of Shemer and Alperovich (2013) who conducted the same laboratory experiment on Peregrine breather solution, showing a better agreement with the simulation of Dysthe equation (higher-order NLS) rather than that of nonlinear Schrödinger equation.

Considering that the above discussion only focuses on the Peregrine breather solution, it cannot be inferred that the NLS equation is generically inaccurate. As reported in the work of Slunyaev et al. (2013), the perfectly stable envelope soliton with very large carrier steepness up to 0.30, i.e. one analytical solution of the NLS equation, has been observed in both numerical simulation and laboratory experiments. Moreover, gradually reducing the carrier wave steepness tends to result in a better and more progressive agreement between simulations and NLS theory. This was recently confirmed by Waseda et al. (2019), who numerically solved the hydrodynamic Euler equation to study the long-term evolution of a water-wave modulated wave train in the optical regime (i.e., very small steepness and spectral bandwidth).

Anyway, the dynamics disclosed by the Peregrine soliton still can shed some light on the formation of abnormal waves. Now, it raises a new question as to whether this kind of breather can exist in the natural ocean environment, which will be considered in the following subsection.

8.1.2 Irregular Background Waves

It is well known that abnormal waves can appear in the ocean wave field, which is usually chaotic. Chabchoub (2016) performed one experimental study, confirming that extreme waves in an irregular JONSWAP wave field can be traced back to originate from the Peregrine breather solution.

The following study uses the same technique and associated parameters in synthesising the initial condition to be consistent with and continue the previous laboratory experiment (Chabchoub 2016). The initial Peregrine soliton is precisely the same as that presented in Fig. 8.1a, i.e., the amplitude of the carrier wave $a = 0.75$ cm and the peak frequency $\omega_0 = 10.68$ rad/s.

The initial irregular background wave is randomly constructed from the retained JONSWAP spectrum. Before the energy partition, the original spectral parameters are the significant wave height $H_s = 3$ cm, the peak frequency $\omega_p = 10.68$ rad/s and the peak enhancement factor $\gamma = 6$. Following the procedure introduced in Fig. 8.5, the hybrid initial wave series can be readily derived. To better understand the flow chart given in Fig. 8.5, the initial discrete amplitude spectra of Peregrine breather solution ($x = -6$ m) and JONSWAP wave (before partition of energy) are illustrated in Fig. 8.6.

One typical example is in Fig. 8.7, where the unstable Peregrine breather packet is visibly embedded in the middle of the irregular background waves (see the bottom). To mimic the randomness in reality, the random phase model is used when constructing the JONSWAP surface elevation, i.e. the amplitudes of wave components are fixed while the phases are randomly selected from the interval [0, 2π). The randomness introduced herein is enough to provide various irregular background waves. It thus leads to a certain

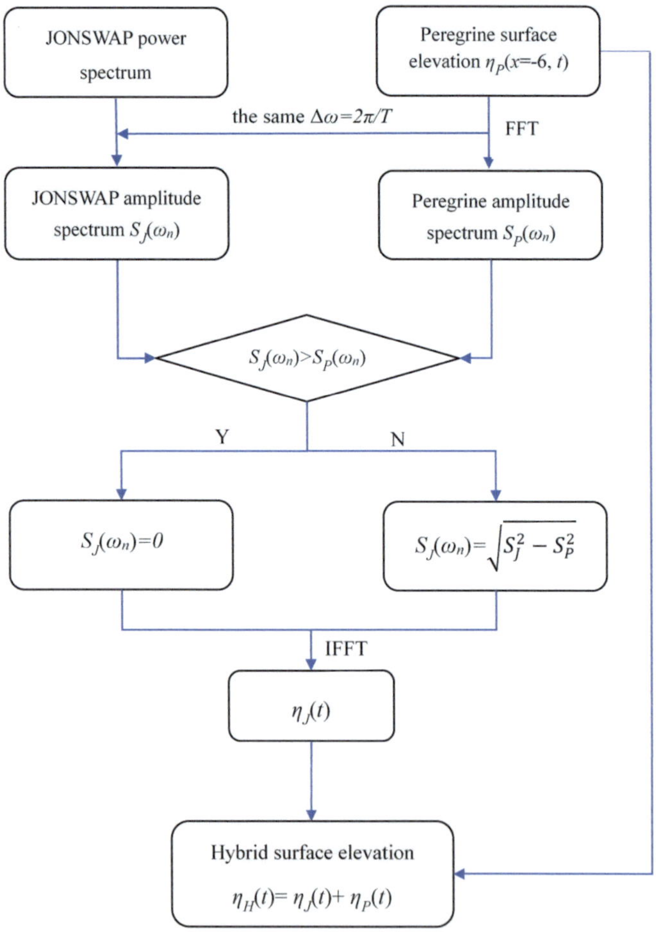

Fig. 8.5 Flow chart of synthesizing the hybrid surface elevation

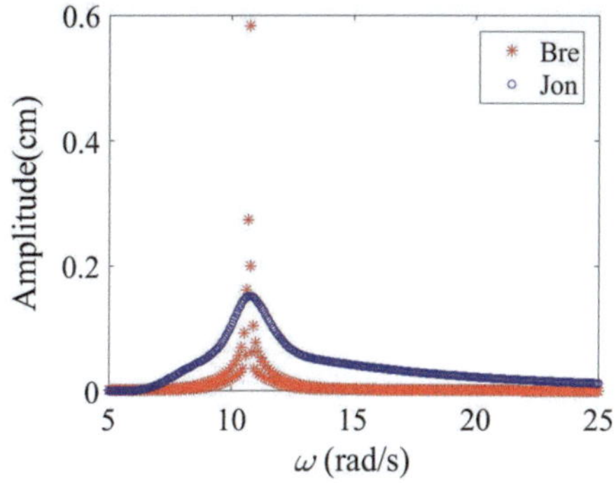

Fig. 8.6 Initial discrete amplitude spectra of Peregrine breather solution ($x = -6$ m) and JONSWAP background wave (before energy partition)

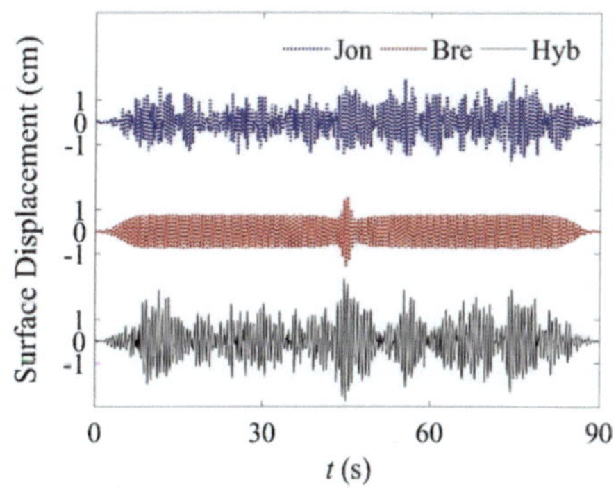

Fig. 8.7 Initial wave series (bottom) synthesized by the linear superposition of Peregrine breather solution (middle) and the irregular background wave (top)

degree of uncertainty on the hybrid initial time series, resulting in different evolutionary tendencies.

Without loss of generality, two representative evolutionary tendencies of the unstable Peregrine packet that is cloaked in the irregular waves are depicted in Fig. 8.8.

As shown in Fig. 8.8a, the initial wave group (0.1 m away from the wavemaker in space) that contains the unstable Peregrine packet, bounded between two vertical red dashed lines, does not present any significant difference from the other initial wave groups at the beginning. Later, an intensive focusing of wave energy can be detected in this group, eventually forming the Peregrine-type extreme wave at the location of 5.3 m, of which the height, calculated with zero-up crossing method, can reach 6.39 cm with the surrounding significant wave height just 2.57 cm.

This Peregrine-type extreme wave is also the maximum wave in the whole wave field. It can be treated as an abnormal or rogue wave considering that its abnormality index, defined as the ratio of maximum wave height to significant wave height, has already exceeded two. In addition, it is very clear that the wave energy piled up to generate the Peregrine-type extreme wave is mainly absorbed from the surrounding waves within the same wave group, being consistent with the phenomenon observed in regular background waves in Fig. 8.2. It has to be stressed that the maximum wave height attained as well as

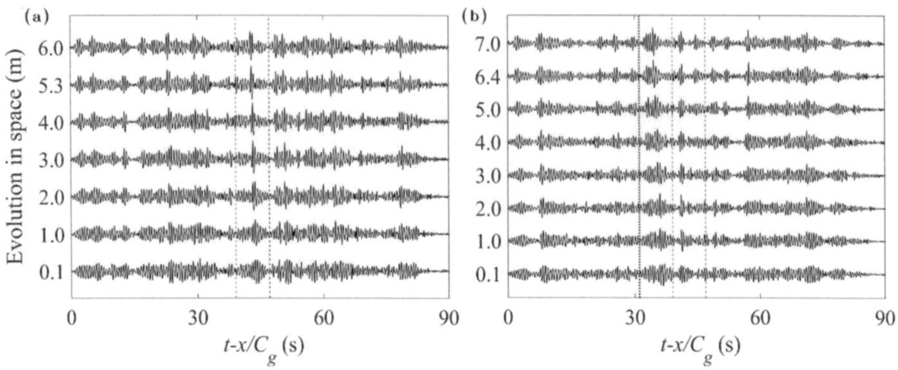

Fig. 8.8 Spatial evolution of the hybrid JONSWAP-Peregrine waves with the specific abnormal wave that appeared ($x = 5.3$ m) in case **a** and disappeared in case **b**. The unstable Peregrine packet is marked between the two red dashed lines. In case **b**, another unexpected rogue wave appears ($x = 6.4$ m) in the group bounded by the blue dotted and red dashed lines

the corresponding evolutionary distance required in this case is quite different from that observed in the laboratory experiment of Chabchoub (2016) because the original irregular background waves cannot be exactly reproduced in this work.

In Fig. 8.8b, the Peregrine-type evolution in the middle of the wave series becomes indiscernible, seemingly due to the less wave energy enclosed in the initial wave group. The irregular background waves have suppressed the modulational instability in the Peregrine breather to a certain degree. Interestingly enough, another extreme wave event appears in the front wave group that is bounded by the blue dotted and red dashed lines. The dynamics involved therein are not clear. However, the formation of abnormal waves is a localized behaviour without significant energy exchange among different wave groups, even in the presence of strong nonlinearity. This property can pave the way for a data-driven prediction scheme for the forecast of extreme waves. That is, the critical elementary wave group (EWG) is identified first from the irregular wave sequence, and the extreme event can be predicted based on the database precomputed by, for example, Dysthe model (Cousins et al. 2019; Farazmand and Sapsis 2017).

Compared to the corresponding evolutionary process in regular background waves, the growth of Peregrine-type waves in irregular sea states seems to be more random. It can be said that the variability is induced by the randomness of initial irregular background waves, which can either enhance or suppress the breather evolution dynamics.

Since the influence of irregular background waves is somewhat uncertain and variable, the evolutionary tendency of Peregrine-type waves must be estimated from a statistical point of view. For a reliable analysis, an ensemble consisting of 1000 cases, randomly synthesized with the Monte Carlo simulation method, has been constructed in this work. To easily distinguish the abnormal wave generated by the Peregrine soliton from that generated by the random wave train alone, only the wave series bounded between 40 and 48s is used for statistical analysis, but the significant wave height is still evaluated based on the entire wave series.

The probability density functions of the maximum wave height and the corresponding evolutionary distance are given in Fig. 8.9, where only the wave groups containing the unstable Peregrine breather packet are considered. Figure 8.9a indicates that the Peregrine-type wave can evolve into or not into the abnormal wave (evaluated by $2H_s$). Moreover, the maximum wave height attained approaches the Gaussian distribution with the mean value close to $2.2H_s$, in line with the central limit theorem. However, in Fig. 8.9b, the probability distribution of the corresponding evolutionary distance, scaled by the theoretical prediction $L_t = 6$ m, is non-Gaussian, and the abscissa covers the entire spatial domain in the numerical simulation. After removing all cases of not generating abnormal waves, the similar probability distribution function can still be derived for the spatial evolutionary distance (not presented herein for brevity). As a result, it will be very difficult to catch the location of extreme events, highlighting the well-known property of "appearing from nowhere" for rogue waves.

More specifically, the chance to observe this unstable Peregrine breather packet, embedded in various irregular background waves, to evolve into an abnormal wave is 62% (see Fig. 8.10a). Nearly two-thirds of these waves can persist in their evolutionary tendencies, corroborating the robustness of breather dynamics in explaining the formation of abnormal waves in irregular sea states.

Since the generation of rogue waves is a localized behaviour, it can be inferred that the maximum crest-to-trough height that can be achieved strongly depends on the energy of the embedded wave group. Extracting all cases with the abnormal wave developed

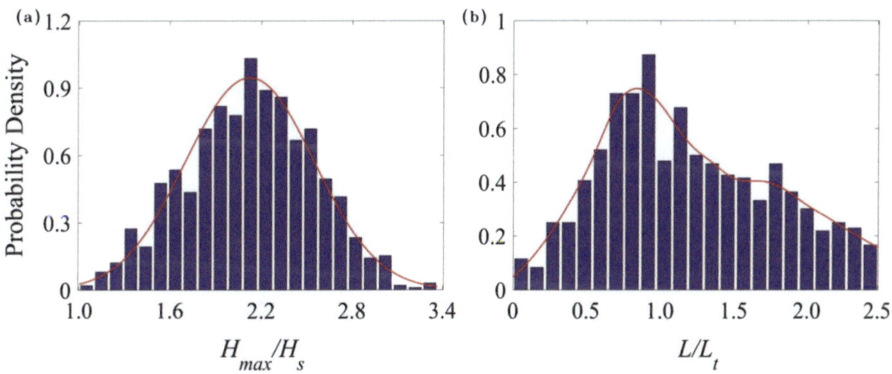

Fig. 8.9 Probability density functions of the maximum wave height **a** attained within the wave group containing the unstable Peregrine breather packet and the corresponding evolutionary distance **b** required to achieve the extreme value. Note that the smooth line is fitted according to the bar diagram

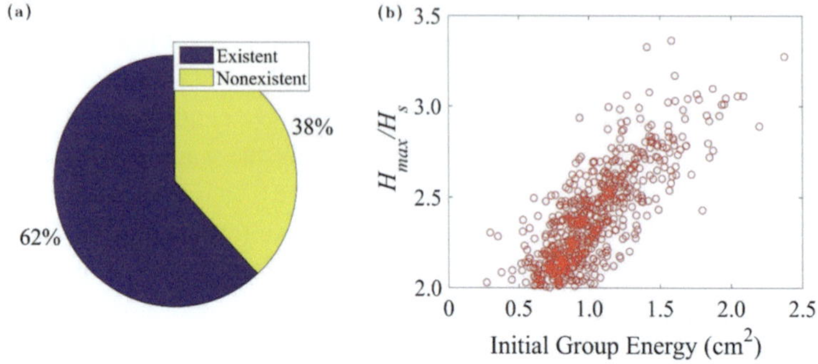

Fig. 8.10 **a** Chart of abnormal waves; **b** the maximum height attained versus the initial wave group energy

by the Peregrine breather dynamics and calculating the initial wave group energy by the variance of the surface elevation (Seiffert and Ducrozet 2018; Seiffert et al. 2017), a linear regression model can be detected in Fig. 8.10b. Generally speaking, the synthesized wave group with more initial energy tends to generate a larger abnormal wave, but the limiting height that can be attained is strongly determined by the detailed structure of the individual group. As revealed by Fig. 8.10b, the difference can be as large as ~ 0.5 H_s for the same initial group energy.

8.2 Influence of Uncertainty in a Stochastic Wave Train

8.2.1 Introduction

It is well known that the profile of the maximum wave identified each time in the same irregular sea state is quite different, which can be attributed to various reasons. Generally speaking, the sources of this uncertainty can be classified into two groups: aleatory (natural) uncertainty and epistemic (knowledge) uncertainty. Aleatory uncertainty represents the intrinsic randomness of the profile of maximum wave in a given sea state, which cannot be reduced or eliminated and thus has to be taken into account in the analysis with the appropriate distribution functions (Bitner-Gregersen 2015).

Epistemic uncertainty indicates the errors in deriving and describing the profile of maximum wave that can be reduced by collecting more information and improving the measuring and modelling methods. According to past research (Bitner-Gregersen 2015; 2017), this type of uncertainty can be further divided into four aspects: data, statistical, model, and climatic.

Data uncertainty is mainly due to the imperfection of an instrument used to measure the maximum surface elevation. Statistical uncertainty is often referred to as estimation uncertainty caused by the limited ensemble size and the estimation technique adopted to evaluate some parameters. Model uncertainty is induced by the assumptions and idealizations made in formulating the physical process. Climatic uncertainty is related to the increasing climatic variability that leads to the deviation from prediction based on past wave records (Bitner-Gregersen et al. 2014).

A lot of methods can be utilized to conduct stochastic modelling to investigate the uncertainty related to maximum waves. For example, the efficient Singular-Vector Perturbation method uses a linear tangent propagator to project small perturbations propagating through a complex nonlinear model. However, the smallness assumption introduced for the perturbation makes it unsuitable for the input wave parameters with large variability. The same problem also exists in the First Order, Second Moment method that has been widely applied to uncertainty analysis in ocean engineering.

In order to improve the accuracy, some more sophisticated methods can be adopted, e.g., the generalized Polynomial Chaos (gPC) method, which expands the random inputs in terms of orthogonal polynomials chosen according to the density functions of the stochastic inputs (Yildirim and Karniadakis 2015). Furthermore, to reduce the computational cost of multi-dimensional random variables, a sparse grid based on Clenshaw-Curtis quadrature can also be used to construct a stochastic collocation simulation.

In fact, the most commonly used method nowadays is the Monte Carlo Simulation which involves randomly sampling a set of values of the basic variables from their probability density functions. It is a computationally demanding method but very straightforward to apply, being capable of providing comprehensive insight into how the combined input uncertainties propagate through a complex model.

Motivated by the randomness of the maximum wave observed in the previous study (Zhang et al. 2016a), the uncertainties induced by the initial conditions are researched herein for the first time. It mainly focuses on the geometrical profile of maximum wave in the space domain, extracted from a large number of stochastic simulations performed with the CS model that has been validated before. A comparison with laboratory observation in the time domain is performed to validate the corresponding numerical analysis.

8.2.2 Case Study

Following the previous work of Zhang et al. (2016a), only two typical sea states (see Table 2.11) characterized by JONSWAP wave spectrum are considered in this subsection (Zhang et al. 2022) because the maximum wave is mainly determined by the ratio of wave steepness to wave spectral width, irrespective of the shape and type of initial wave spectrum as illustrated in Shemer et al. (2010b).

To simulate the irregular water waves characterized by the JONSWAP power spectrum, the initial surface elevation $\eta(x)$ is assigned in the following general form

$$\eta = \sum_{j=1}^{J} a_j \cos(k_j x + \theta_j), \tag{8.1}$$

where, for each wave component j, a_j is a random (over different runs) amplitude with Rayleigh distribution and θ_j is a random phase uniformly distributed in the interval $[0, 2\pi)$. The expected value of the square of each wave amplitude can be related to the wavenumber spectrum by

$$\frac{1}{2}E[a_j^2] = S(k_j)\Delta k, \tag{8.2}$$

where $S(k)$ it is the same as that presented in Eq. (7.24).

In the numerical simulation, the duration of each run is $57T_{p,}$ and the computational spatial domain is around $60L_p$ where L_p denotes the peak wavelength. For these two sea states, the total number of initial free wave components J are 150 and 120, respectively. Note that some of these parameters will be modified slightly in the later numerical simulation to evaluate their influence on the uncertainty of the maximum wave profile.

The continuous form of the initial JONSWAP wave spectrum is shown in Fig. 8.11, where the wavenumber has been nondimensionalized by the peak wavenumber. The initial surface elevation can be synthesized with the linear superposition method. However, to better reproduce the property of uncertainty, the random-phase/amplitude model is usually utilized to generate more diverse initial conditions.

After a screening of Eq. (8.1) and the numerical scheme in the CS model, three types of uncertainties in the initial conditions can be identified, mainly attributed to the following input parameters: the random amplitude a_j and the random phase θ_j (natural uncertainty), the wavenumber k_j and the total number of free wave components J (knowledge uncertainty), and the length of spatial wave train in the computational domain (mixed uncertainty). Their variations are quite different and thus can contribute significantly to the variability and uncertainty of maximum waves.

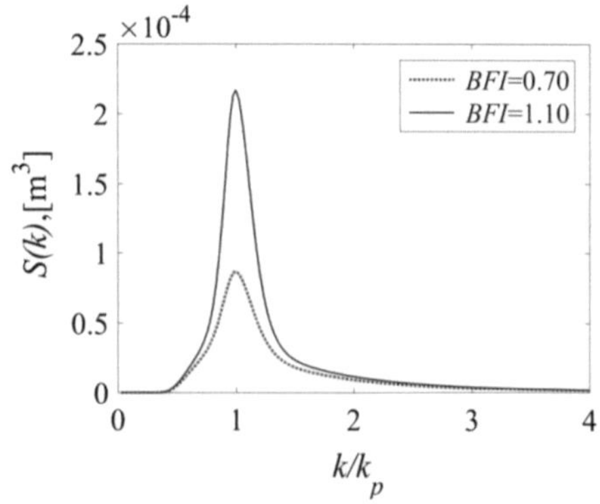

Fig. 8.11 Initial wavenumber spectra of the two typical sea states

Moreover, it can be further inferred that some of these parameters are not independent. Some techniques, such as Rosenblatt or Nataf transformation could be applied to transform the dependent non-normal variables into independent standard normal variables, but their joint density function or conditional distribution (not available herein) must be given first. Meanwhile, the convenient method, such as the First Order Second Moment method, is no longer valid due to the significant covariance and considerable uncertainty.

Therefore, the Monte Carlo Simulation method is adopted in this study to investigate the uncertainty of maximum wave induced by these initial input parameters. To indicate their effects clearly and reduce the computational cost, the sources of uncertainty will be addressed separately for each sea state with the following three steps:

I. Conduct sensitivity and uncertainty analysis based on 2000 random runs, of which the initial conditions are synthesized with the random-phase/amplitude model (i.e., a_j and θ_j are variables while the other three types of input parameters are fixed);
II. Randomly sampling 1000 groups of amplitudes and phases, which are fixed for the initial condition, calculate the bias of maximum wave, from a statistical point of view, arising from the uncertainty of k_j and J, respectively;
III. Select five typical cases to study the uncertainty induced by the length of a wave train in the computational domain (in x direction), which is the combined effects of the above four types of input parameters.

8.2.3 Definition of Maximum Waves

For irregular waves, two definitions of individual waves are frequently used in the wave-by-wave analysis: zero-down-crossing and zero-up-crossing, which could impose various degrees of difference in the statistical results even for the same wave record. To exactly extract the profile of the maximum wave, i.e. the highest one from the whole numerical wave field, both zero-up-crossing and zero-down-crossing methods are used to analyse the same realization and only the one with the largest crest-to-trough height is retained, thus leading to a sample of 2000 individual waves (data from step I in Sect. 8.2.2).

According to the shapes of the wave profile, maximum waves in these two sea states can be categorized into four groups with the corresponding percentage given in Fig. 8.12. The classification is determined by two factors: which part (crest or trough) dominates the wave height and where (front or rear) the dominant part is located. Maximum waves tend to appear in the form of a huge crest in the front part (FrontPos), which is particularly pronounced (57%) in the case with larger *BFI*. It is consistent with the previous conclusion in Slunyaev et al. (2016), and this characteristic is also valid for the directional maximum wave, provided the wave spectrum (both in frequency and direction) is narrowband (Fujimoto et al. 2018).

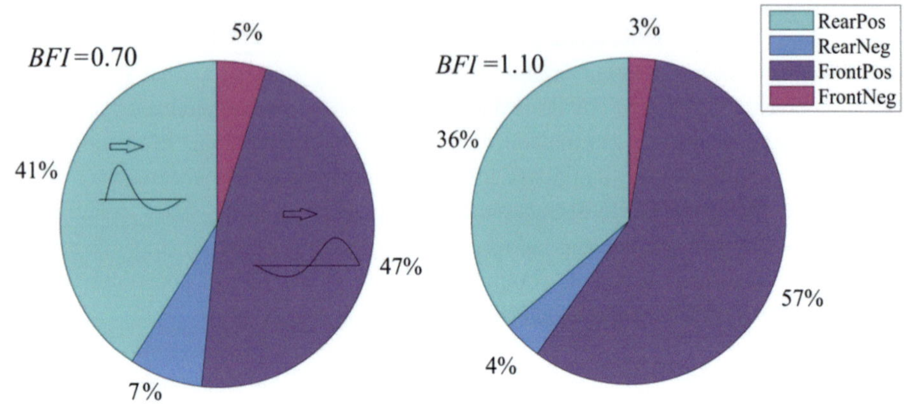

Fig. 8.12 Charts of maximum waves classified into four types. The characteristic wave shapes are plotted in the first chart, where the arrows indicate the propagation direction in space. 'Pos' and 'Neg' mean that the maximum wave mainly comprises wave crest and trough, respectively. 'Rear' and 'Front' illustrate the location of the significant part of the maximum wave

Moreover, regardless of the location of the wave crest, the proportion of maximum waves with a huge crest is much higher, i.e. 88 and 93% (RearPos + FrontPos) in the two sea states, respectively. Thus, it is reasonable to define the basic profile of the maximum wave with the centre on the large wave crest, as shown in Fig. 8.13.

Interestingly enough, if the nonlinearity is not very strong, i.e. $BFI = 0.70$ in this study, the ratio of zero-up-crossing (RearPos + FrontNeg) to zero-down-crossing (FrontPos + RearNeg) maximum waves is 46:54. As the Benjamin-Feir index increases, this ratio decreases to 39:61, implying that with zero-down-crossing method more large waves can be detected in statistical analysis in the presence of strong nonlinearity. However, the proportion of maximum waves extracted with the zero-up-crossing method is still very

Fig. 8.13 Sketch illustrating the parameters used to describe the basic profile of maximum wave

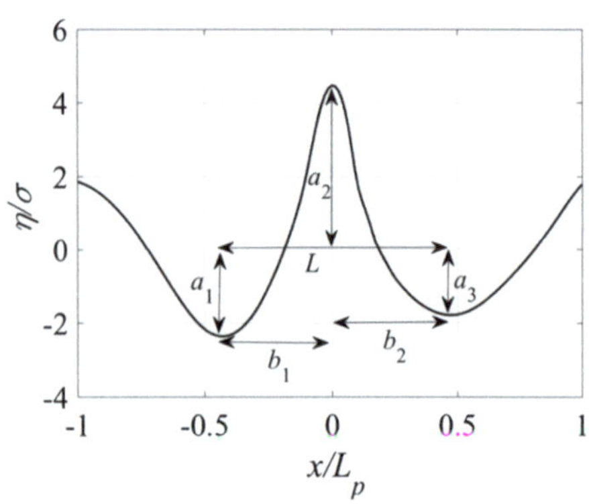

high. Hence, to catch the uncertainty property in the profile of maximum waves, both zero-up-crossing and zero-down-crossing definitions must be utilized in the later analysis.

8.2.4 Sensitivity Analysis

The formation of maximum wave and the subsequent possible steepness-induced wave breaking is a highly localised behaviour. To better evaluate this kind of properties, a comprehensive description of the profile of maximum wave is very important. Therefore, six parameters similar to those defined in (Babanin et al. 2010) are introduced in this work (see Fig. 8.13), involving the horizontal asymmetry parameter *HA*, the vertical asymmetry parameters VA_1 and VA_2, the length of maximum wave *L*, the maximum crest-to-trough height *H* and the fatness coefficient C_b.

$$HA = b_1/b_2, \tag{8.3}$$

$$VA_1 = a_2/a_1, \tag{8.4}$$

$$VA_2 = a_2/a_3, \tag{8.5}$$

$$L = b_1 + b_2, \tag{8.6}$$

$$H = a_2 + \max(a_1, a_3), \tag{8.7}$$

$$C_b = A/(LH), \tag{8.8}$$

where a_1 and a_3 are the absolute depth of the back and front troughs in sequence, and a_2 is the elevation of the crest, b_1 and b_2 denote the horizontal distance in space from the crest to the back and front troughs, respectively. The length of maximum wave *L* is now the horizontal trough-to-trough distance, and the formulation of maximum crest-to-trough height *H* in Eq. (8.7) considers the influence of the individual wave definition discussed in the above section.

To indicate the overall condition of the maximum wave, fatness coefficient C_b is also introduced in Eq. (8.8) where *A* is the area under the maximum surface wave integrated from the back trough to the front one, with the bottom horizontal line defined to pass through the lowest point of the deeper trough. In Fig. 8.13, L_p and σ represent the corresponding surface elevation's peak wavelength and standard deviation, respectively. Besides, the heights of maximum wave crest *C* and wave trough *T* in the same sea state are also discussed in the following analysis because they may be more concerned by some engineers and researchers, and the sum of these two parameters can be used to compare with the height of maximum wave to explore the evolutionary characteristics. Note that the information associated with the crest and trough of the maximum wave is not directly given but implicitly indicated by Eqs. (8.4), (8.5) and (8.7).

The random-phase/amplitude model is the most critical part in synthesizing the initial condition for the numerical simulation of irregular waves (Holthuijsen 2007). It involves $2J$ random input parameters, capable of reasonably reproducing the natural uncertainty associated with the profile of maximum wave. To better evaluate the impact of variation in the random-phase/amplitude model on the properties of the maximum wave, sensitivity analysis is performed first for the given two sea states.

Among so many different methods of sensitivity analysis, Monte Carlo Simulation (statistical method) is adopted in this study because it is a global variance-analysis-based method and can exhibit the contribution of each variable to the variance in the output (Bertrand and Saltelli 2017). The sensitivity parameter is defined as below

$$\beta_j = b_j^2 \sigma_j^2 / \sigma_Y^2, \tag{8.9}$$

where b_j is the coefficient of regression model between the initial variable such as amplitude a_j and the specific output parameter Y defined above to evaluate the characteristics of maximum waves. σ_j and σ_Y are the corresponding standard deviations, respectively.

The sensitivity of maximum crest-to-trough height to the initial wave amplitudes is given in Fig. 8.14. The most sensitive part is around the peak wavenumber, i.e. about $0.8k_p \sim 1.2k_p$. Moreover, it can be found that the maximum crest and trough heights also present a similar sensitivity to the initial amplitudes. Thus, on the one side, it reveals the dominant mechanism to generate the maximum crest, trough and crest-to-trough heights should be the same; on the other side, to a certain degree, it can confirm the previous conclusion that in unidirectional water waves, the formation of rogue or abnormal waves is mainly due to the quasi-resonant quadruplet wave-wave interaction, which occurs primarily in the peak frequency domain.

To be concise, the other figures related to sensitivity analysis are not shown herein. In summary, for the maximum crest, trough and crest-to-trough heights, their sensitivities to the initial random phases are less significant than those to the initial random amplitudes. Moreover, the other five output parameters show very similar low order sensitivity to the initial amplitudes and phases, independent of the wavenumber and sea state.

In other words, for maximum wave, except for the crest-to-trough height, all wave components almost equally determine the other geometrical properties. As a result, no wave component can be ignored in the uncertainty analysis below. It has to be mentioned in the sensitivity analysis that the initial random phases and amplitudes are used simultaneously in the numerical simulation because random phases do not significantly affect the final results.

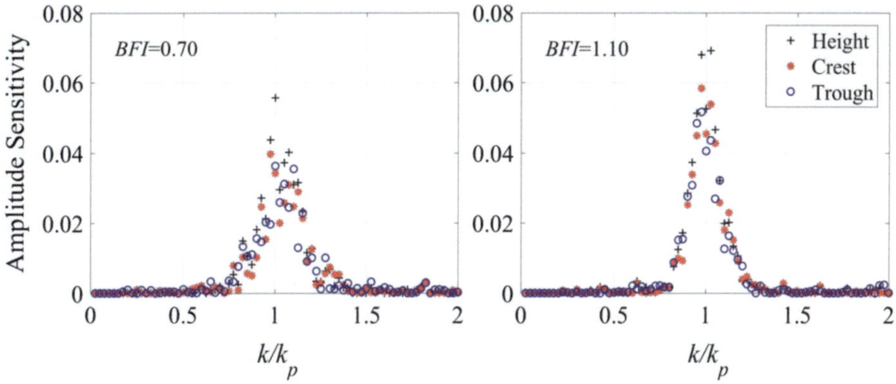

Fig. 8.14 Sensitivities of the maximum crest, trough and crest-to-trough heights to the initial wave amplitudes

8.2.5 Aleatory Uncertainty

Due to the aleatory (natural) uncertainty, statistical quantities have to be used to describe the profile of the maximum wave. Following the above sensitivity analysis, the mean value and standard deviation of all output parameters related to maximum wave and maximum crest and trough are listed in Table 8.1 for the two typical sea states.

The expected value of maximum crest-to-trough height in both sea states is around 6.955σ. Combined with the corresponding probability density function displayed in Fig. 8.15, it is very clear that the uncertainty is decreased (narrower distribution) with a larger Benjamin-Feir index. The same tendency can also be detected for the maximum crest and trough heights (not presented herein). Moreover, it can be further inferred that in most cases, the maximum wave crest and trough cannot appear simultaneously to form a huge wave because the sum of their mean values is much larger than the expected value of maximum crest-to-trough height, in agreement with the former conclusions drawn from Peregrine breather wave (Slunyaev and Shrira 2013).

Table 8.1 Aleatory uncertainty in the profile of maximum wave as well as the heights of maximum crest and trough

BFI	Item	H/σ	C/σ	T/σ	L/L_p	*HA*	VA_1	VA_2	C_b
0.70	Mean	6.95	4.39	3.48	0.93	1.19	2.12	2.36	0.25
	Std. Dev	0.63	0.49	0.32	0.17	0.52	0.87	1.25	0.03
1.10	Mean	6.96	4.54	3.39	0.95	1.36	2.05	2.71	0.25
	Std. Dev	0.52	0.42	0.28	0.20	0.52	0.67	1.42	0.03

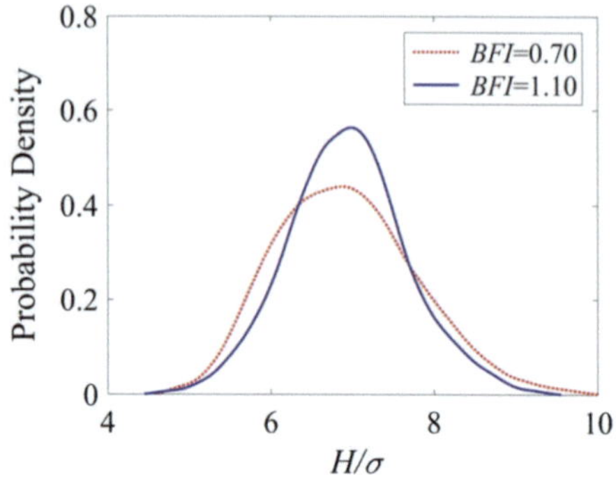

Fig. 8.15 Probability density function of the crest-to-trough height of maximum wave

Regarding the length of maximum waves, the most probable value is close to $0.94L_p$, being contracted in comparison with the initial peak wavelength, and these waves tend to lean forward, considering the horizontal asymmetry parameter *HA* is larger than one (Babanin et al. 2007). Since the mean value of VA_1 is smaller than that of VA_2, it can be concluded that a_1 is larger than a_3 (see Fig. 8.13), meaning that the back wave trough is inclined to be deeper than the front one.

To save space, only two representative probability density functions (pdf) are exhibited in Fig. 8.16. The pdf of the length of maximum wave (L), plotted in the left panel of Fig. 8.16, is very similar to those of the asymmetry parameters (*HA*, VA_1 and VA_2), and the only difference is in the length of the right tail, illustrated by the various standard deviations in Table 8.1. The large uncertainties of these asymmetry parameters imply the difficulty in determining the profile of maximum waves. The distribution of fatness coefficient C_b is unique compared to the other output parameters. Although the corresponding statistical quantities in Table 8.1 are the same for both sea states, their probability density functions in Fig. 8.16 are slightly different, i.e., narrower (less uncertain) in the case of larger *BFI*.

8.2.6 Epistemic Uncertainty

To synthesize the initial surface elevation, normally, the continuous wave spectrum in Fig. 8.11 has to be discretized and transformed into the discrete amplitude spectrum with all wavenumbers k_j in Eq. (8.1) determined in this process. Theoretically speaking, k_j could be any value in the corresponding wavenumber interval. However, due to the lateral periodic boundary condition imposed in the numerical wave model, the wavenumber resolution must be constant over the entire range, i.e. $\Delta k = k_{j+1} - k_j$ is the same for all j. Therefore, to check the uncertainty induced by wavenumbers, all k_j can only be shifted together to the left or right in the wavenumber domain with a maximum magnitude of Δk.

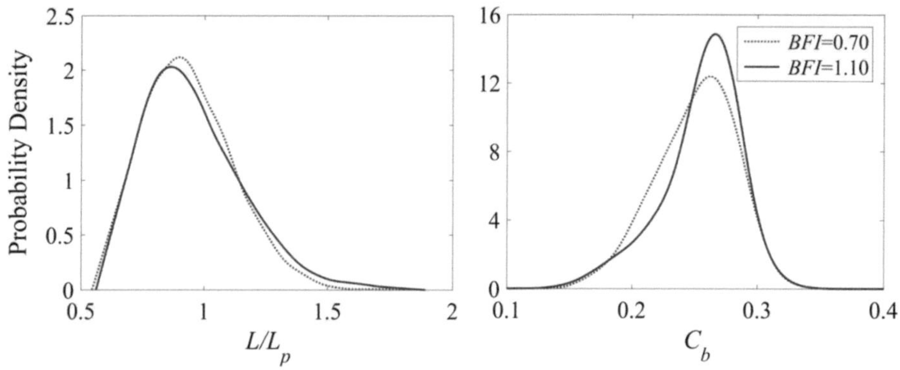

Fig. 8.16 Probability density functions of the length of maximum wave and the block coefficient

To a certain extent, the total number of free waves J is also assigned subjectively. With the wavenumber resolution Δk fixed, a slight variation of J (indicated by the change of wave energy, less than 4%) is also performed in this section to assess the corresponding uncertainty.

The original profile of the maximum wave is modified randomly when the wavenumbers and the number of free waves are changed, strongly depending on the initial amplitude and phase of each wave component. To give an intuitive feeling on the effect of modification, two typical examples are first shown in Fig. 8.17, where the left figure is due to the shift (left and right) of all wavenumbers with a magnitude of Δk (equal to the wavenumber resolution) and the right one is the result of variation (decrease and increase) of the number of free waves around the high cut-off frequency with a size of 6.

After modification of the wavenumber or the number of free waves, the maximum height that can be achieved may not be from the same individual wave in the new run, and even if from the same one, the space and time where its maximum height appears can be different from those in the original run sometimes. Therefore, it is very difficult to trace its variation, and a compromised solution is given: extract the wave profile from the new run at the space and time where the maximum wave appeared in the original run and then compare it with the original maximum wave profile.

To better quantify these uncertainties, statistical quantities associated with the bias (B) are calculated for all output parameters.

$$B_i = (m_i - o_i) \big/ o_i, \tag{8.10}$$

where o_i is the original output parameter, and m_i is the corresponding modified output parameter in the same realization i. Note that different realizations have different initial amplitudes and phases randomly selected, as stated in Step II (see Sect. 8.2.2).

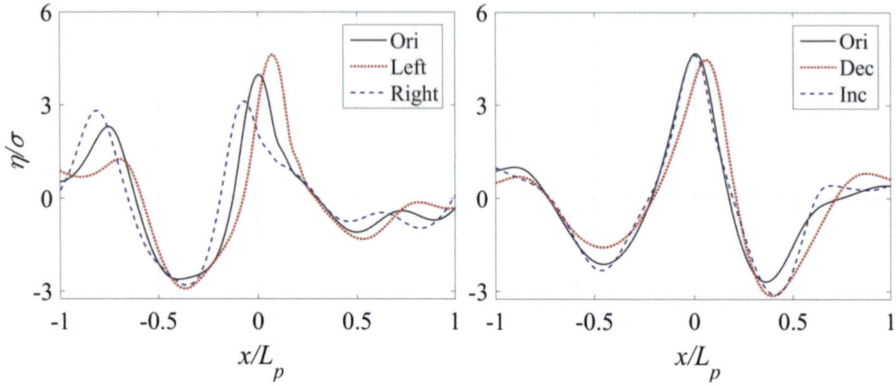

Fig. 8.17 Example of modification on the profile of maximum wave induced by uncertainties in wavenumber and the number of free waves. 'Ori' denotes the original case. 'Left' and 'Right' represent the cases with all initial wavenumbers shifted to the left and right. 'Dec' and 'Inc' indicate the cases with the total number of free waves J decreased and increased separately

The mean value and standard deviation of bias of all output parameters are summarized in Tables 8.2 and 8.3, respectively. According to the magnitudes of these statistical quantities, the uncertainties of these parameters can be roughly ranked into three grades: high, moderate and low.

The high uncertainty is associated with the three asymmetry parameters (HA, VA_1 and VA_2), meaning that the vertical and horizontal asymmetries of maximum waves are very sensitive to the initial wavenumbers and the number of free waves. The moderate uncertainty is related to the length of maximum wave L and the fatness coefficient C_b.

Table 8.2 Uncertainty induced by wavenumbers

BFI	Shift	Item	H/σ	C/σ	T/σ	L/L_p	HA	VA_1	VA_2	C_b
0.70	Left	Mean	0.23%	0.24%	0.88%	6.75%	20.30%	15.18%	24.92%	– 0.48%
		Std. Dev	0.06	0.12	0.09	0.26	0.82	0.75	0.96	0.17
	Right	Mean	– 0.07%	– 1.20%	– 0.61%	0.95%	23.24%	15.15%	30.68%	1.35%
		Std. Dev	0.05	0.10	0.10	0.24	0.78	0.67	1.05	0.16
1.10	Left	Mean	0.70%	1.76%	0.85%	6.13%	17.71%	14.72%	19.94%	1.46%
		Std. Dev	0.05	0.10	0.08	0.28	0.76	0.56	1.03	0.20
	Right	Mean	– 0.63%	– 0.50%	– 0.85%	– 0.71%	18.03%	10.15%	30.26%	3.60%
		Std. Dev	0.05	0.10	0.08	0.27	0.72	0.54	1.09	0.19

Table 8.3 Uncertainty induced by the total number of free waves

BFI	Variation	Item	H/σ	C/σ	T/σ	L/L_p	HA	VA_1	VA_2	C_b
0.70	Decrease	Mean	– 1.11%	– 0.23%	– 0.59%	0.15%	10.13%	5.73%	10.35%	1.61%
		Std. Dev	0.03	0.05	0.04	0.18	0.54	0.40	0.51	0.12
	Increase	Mean	1.44%	0.70%	1.53%	2.25%	11.15%	1.71%	6.84%	– 0.64%
		Std. Dev	0.03	0.04	0.04	0.19	0.51	0.37	0.57	0.12
1.10	Decrease	Mean	– 0.63%	– 0.16%	– 0.97%	2.12%	6.63%	7.31%	9.67%	1.56%
		Std. Dev	0.03	0.05	0.04	0.25	0.55	0.39	0.69	0.17
	Increase	Mean	0.79%	0.66%	0.76%	4.49%	5.63%	7.41%	4.74%	0.23%
		Std. Dev	0.03	0.05	0.04	0.25	0.48	0.40	0.71	0.16

Their uncertainties are relatively low for the rest of output parameters, i.e., the maximum crest, trough and crest-to-trough heights.

Due to the central limit theorem, the probability density function of the bias approaches Gaussian distribution for all output parameters, irrespective of sea state. One representative example is given in Fig. 8.18 for the bias of the maximum crest-to-trough height. Specifically, if all wavenumbers are shifted to the left side by Δk, the most probable bias will be positive (i.e., higher waves) and vice versa. However, this is not true, as revealed by the bias span in the first picture of Fig. 8.18. The same argument can apply to the case of changing the total number of free waves. If the wave energy is slightly increased (less than 4%), the most probable effect is the increase in maximum wave height. Despite the small variation in energy, it is still possible to generate a large discrepancy in the magnitude of maximum crest-to-trough height.

A similar conclusion can also be drawn regarding the other output parameters, but it will not be discussed in detail. If the output, such as the asymmetry parameter, is equally sensitive to all initial wave components, the variability induced by small epistemic uncertainty in the initial condition can be considerably large.

8.2.7 Mixed Uncertainty

This section considers the uncertainty caused by the length of the wave train in the wave propagation direction. To a certain degree, it is equivalent to the combined randomness of

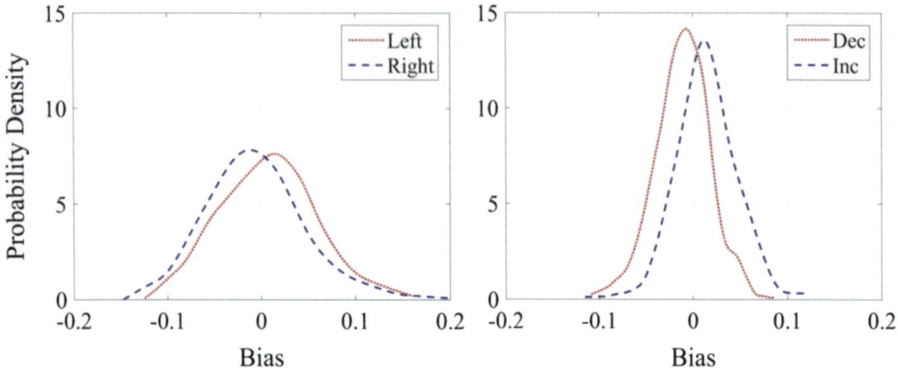

Fig. 8.18 Probability density function of the bias of maximum crest-to-trough height. The legends are the same as in Fig. 8.17

the above four types of input parameters because the length of the wave train in the computational domain determines the wavenumber resolution Δk, which is used to discretize the continuous wave spectrum. Once the spectral resolution is changed, the other initial parameters will be modified in the following procedure to synthesize the initial condition.

In fact, for a given sea state, a variation in the wave train length implies a change in the number of individual waves within one modulation period in the space domain. For the fully nonlinear CS model, this can be realized by modifying the location of the peak wavenumber, denoted by N, in the wavenumber domain. In Fig. 8.19, only five different lengths are considered, and the corresponding error bars are presented for those typical output parameters.

With the increased length of the wave train (larger N), the maximum crest-to-trough height, as well as the maximum trough height (not presented herein), is almost invariant while the maximum crest height is gradually decreased. Considering that the locations of maximum crest height and crest-to-trough height are very close, it can be inferred that the crest elevation of the maximum wave is correspondingly decreased. Therefore, the invariance of maximum crest-to-trough height means that the depth of the corresponding trough (see Fig. 8.13) must be increased. This conclusion is also indirectly confirmed by the decreased tendency of vertical asymmetry parameter VA_1, which is in agreement with the changing trend of VA_2. Meanwhile, the maximum wave will become fatter, as revealed by the increased fatness coefficient C_b. Last, it seems that the length of the maximum wave and the horizontal asymmetry are not sensitive to the length of the wave train.

In order to verify and validate the above numerical analysis, time series are extracted from the numerical simulation with the method in (Slunyaev et al. 2014) and compared with the corresponding laboratory observation (Onorato et al. 2009; Toffoli et al.

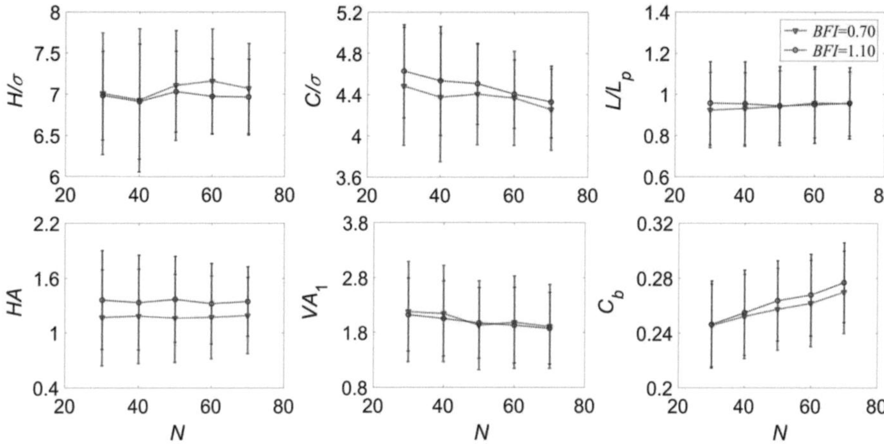

Fig. 8.19 Variation on the profile of maximum wave induced by uncertainty in the length of wave train

2010a). The physical wave tank is 50m, which is long enough for waves to evolve due to modulational instability.

The profiles of maximum wave in the time domain defined with zero-up-crossing (deeper rear trough) and zero-down-crossing (deeper front trough) methods are displayed in Figs. 8.20 and 8.21, respectively. Generally speaking, the maximum wave detected in numerical simulation is in conformity with that observed in laboratory experiments in both sea states. The large width of the shaded error bar implies that the uncertainty on the profile of the maximum wave is very high, which definitely will result in a large variation on the asymmetric parameters defined to describe the property of the maximum wave. Following the conclusion derived in the analysis of space series, more maximum waves tend to be extracted from time series with zero-up-crossing method (i.e., deeper rear trough).

Actually, this can also be confirmed by abnormal waves such as Andrea, Camille and Draupner waves, which are occasionally derived from field measurements in the open sea (Guedes Soares et al. 2004; Fedele et al. 2017). It can be argued again that the nature and knowledge uncertainties make it very difficult to predict the profile of maximum waves even in a given sea state.

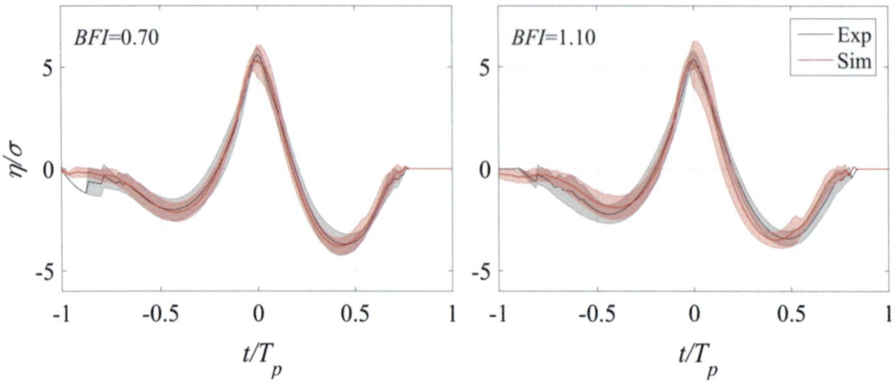

Fig. 8.20 Profile of maximum wave defined with zero-up-crossing method

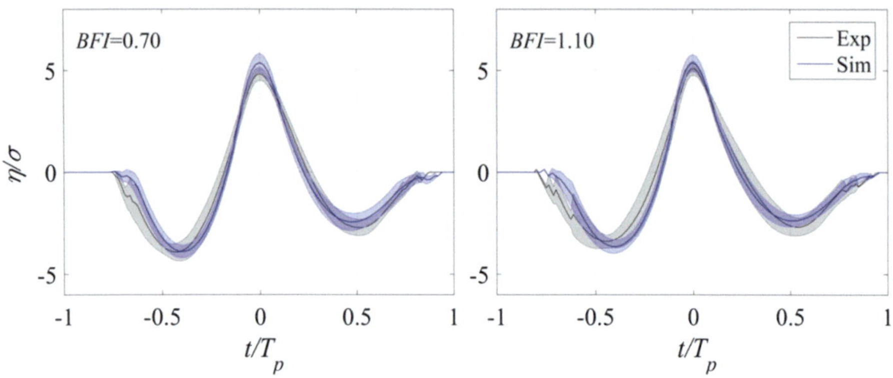

Fig. 8.21 Profile of maximum wave defined with zero-down-crossing method

8.3 Conclusion

As the prototype of an extreme wave event, the evolutionary tendencies of the unstable Peregrine breather packet are studied in the first section of this chapter, not only in the regular background waves but also in the 2D irregular background waves. Only the maximum amplification factor and the corresponding evolutionary distance are fully discussed herein because these two variables are most useful in engineering applications, especially when conducting laboratory experiments related to the transient response of floating structures to abnormal waves.

In the focusing process with the regular background waves, the Peregrine breather packet contracts considerably and absorbs energy from the surrounding environment, mainly from the back waves. This energy transfer process is partially irreversible. Due to wave energy conservation, the growing processes of the maximal crest and trough amplitudes within the unstable breather packet are out of phase. Consequently, the abnormal wave cannot simultaneously appear with the highest crest and deepest trough.

In contrast to the NLS theory, the individual modulation period of wave amplitude is enlarged in the fully nonlinear simulation, thus elongating the whole evolutionary process. The increased crest and decreased trough amplitudes reduce the smoothness of the growing process of the maximum wave envelope.

The chaotic background waves characterized by the JONSWAP wave spectrum significantly influence the evolutionary tendency of the unstable Peregrine breather packet. From a statistical point of view, the breather dynamics considered herein can persist in most cases, substantiating its physical robustness to external perturbations. However, the evolutionary speed is changed randomly, and the spatial location of maximum wave height becomes arbitrary, making the abnormal wave appear from nowhere.

The maximum crest-to-trough height attained by the Peregrine-type wave packet in irregular background waves is mainly determined by the energy of the surrounding waves, which generally belong to the same wave group. No significant interaction with other wave groups can be detected. Since the formation of abnormal waves is a localized behaviour, it is possible to predict the extreme event by identifying the critical elementary wave groups and calculating their initial energy. However, to get the exact amplification factor is still challenging because it also depends on the detailed structure of the individual wave group.

Note that the conclusions derived in this chapter is only limited to one specific case. The probability of focusing-persistence should be determined by the particular combination of the phases and the relative strength between the breather solution and irregular background wave. To get a comprehensive and complete conclusion on the evolutionary property of Peregrine-type wave packet embedded in irregular background waves, a series of different combinations between them should be further considered.

The randomness of maximum waves in two typical short-term sea states is also considered in the second part of this chapter. The main concern herein is the uncertainty induced by the initial condition, which is evaluated using six output parameters related to the profile of the maximum wave. Meanwhile, the uncertainties associated with the maximum crest and trough heights are also presented.

Different definitions of individual waves can give rise to different maximum waves in analysing the same wave record. As revealed in this chapter, with zero-up-crossing and zero-down-crossing methods, the probability of detecting the maximum wave is almost equal in the case of small *BFI*. If the nonlinearity is increased, i.e. larger *BFI*, it has more opportunities to detect the maximum wave with the zero-down-crossing method (deeper rear trough). Despite different zero-crossing methods, the vast majority of maximum waves detected herein are mainly composed of a large wave crest.

The sensitivity analysis indicates that the maximum crest, trough and crest-to-trough heights are very sensitive to the amplitudes of some specific wave components (around the peak wavenumber in the wavenumber domain). At the same time, the other output parameters do not show a similar correlation. For all output parameters concerned in this work, their sensitivities to the initial random phases are lower than those to the initial random amplitudes, and the phase of each wave component seems to affect the profile of the maximum wave equally.

The randomness caused by the initial random-phase/amplitude model (natural uncertainty) is very large, demonstrated by the high variability and uncertainty of all output parameters. Surprisingly enough, modulational instability indicated by the Benjamin-Feir index can reduce part of the uncertainties related to the crest-to-trough heights of maximum waves. In addition, most maximum waves tend to lean forward with the back trough deeper than the front one and the trough-to-trough distance less than the peak wavelength.

The uncertainties induced by the wavenumber and the number of free waves (knowledge uncertainty) are illustrated by the mean value and standard deviation of the bias of all output parameters. In both sea states, the behaviours of the three asymmetry parameters are highly uncertain, while the maximum crest, trough and crest-to-trough heights seem to be least influenced. As a result of the central limit theorem, the probability density functions of the bias are all approximately normally distributed, irrespective of sea state. Although some general trends can be observed, the effect of modification on the profile of the maximum wave is random, strongly depending on the initial amplitude and phase of each wave component.

The uncertainty generated by the length of the spatial wave train is complex because it represents the combined randomness of all the above input parameters (mixed uncertainty). With the increased number of individual waves in the computational domain, the crest height of the maximum wave gradually decreases, but the depth of its trough tends to increase, making the maximum crest-to-trough height nearly constant. Meanwhile, the maximum wave is inclined to become fatter while its length and horizontal asymmetry are almost invariant.

It has to be admitted that only the initial condition uncertainty is highlighted in this chapter. There are a lot of other factors that can contribute greatly to the uncertainty of the profile of the maximum wave. For example, the inaccurate numerical scheme can lead to numerical dissipation and phase errors, which can smear out the amplitude of the maximum wave and shift its position. Therefore, a complete and comprehensive study is needed in the future.

References

Akhmediev, N., Eleonskii, V., Kulagin, N., 1985. Generation of periodic trains of picosecond pulses in an optical fiber: exact solutions. Soviet Physics, JETP 62, 894–899.

Alberello, A., Chabchoub, A., Monty, J.P., Nelli, F., Lee, J.H., Elsnab, J., Toffoli, A., 2018. An experimental comparison of velocities underneath focussed breaking waves. Ocean Engineering 155, 201–210.

Annenkov, S., Shrira, V., 2009. Evolution of kurtosis for wind waves. Geophysical Research Letters 36(13), L13606.

Antao, E., Guedes Soares, C., 2016. Approximation of the joint probability density of steepness and height of individual waves with a bivariate gamma distribution. Ocean Engineering 126, 402–410.

Arena, F., Fedele, F., 2002. A family of narrow-band nonlinear stochastic processes for the mechanics of sea waves. European Journal of Mechanics B/Fluids 21, 125–137.

Babanin, A.V., Chalikov, D., Young, I.R., Savelyev, I., 2007. Predicting the breaking onset of surface water waves. Geophysical Research Letter 34(7), L07605.

Babanin, A.V., Chalikov, D., Young, I.R., Savelyev, I., 2010. Numerical and laboratory investigation of breaking of steep two-dimensional waves in deep water. Journal of Fluid Mechanics 644, 433–463.

Babanin, A.V., Onorato, M., Cavaleri, L., 2019. On natural modulational bandwidth of deep-water surface waves. Fluids 4(2), 67.

Baldock, T.E., Swan, C., Taylor, P.H., 1996. A laboratory study of nonlinear surface waves on water. Philosophical Transactions of the Royal Society A 354: 649–676.

Bateman, W.J.D., Katsardi, V., Swan, C., 2012. Extreme ocean waves. Part I. The practical application of fully nonlinear wave modelling. Applied Ocean Research 34, 209–224.

Bertrand, L., Saltelli, A., 2017. Introduction to Sensitivity Analysis. In: Ghanem, R., Higdon, D., Owhadi, H. (eds) Handbook of Uncertainty Quantification. Springer, Cham. 1103–1122.

Bitner-Gregersen, E.M., 2015. Joint met-ocean description for design and operations of marine structures. Applied Ocean Research 51, 279–292.

Bitner-Gregersen, E.M., 2017. Rethinking rogue waves. DNV GL R&I Position Paper, 1–7.

Bitner-Gregersen, E.M., Toffoli, A., 2012. On the probability of occurrence of rogue waves. Natural Hazards and Earth System Science 12, 751–762.

Bitner-Gregersen, E.M., Bhattacharya, S.K., Chatjigeorgiou, I.K., Eames, I., Ellermann, K., Ewans, K., Hermanski, G., Johnson, M.C., Ma, N., Maisondieu, C., Nilva, A., Rychlik, I., Waseda, T., 2014. Recent developments of ocean environmental description with focus on uncertainties. Ocean Engineering 86, 26–46.

H. Zhang and C. Guedes Soares, *Numerical Modelling of Extreme Waves*, Synthesis Lectures on Ocean Systems Engineering, https://doi.org/10.1007/978-3-031-77084-5

Bitner-Gregersen, E.M., Gramstad, O., Magnusson, A.K., Malila, M.P., 2021. Extreme wave events and sampling variability. Ocean Dynamics 71, 81–95.

Bitner-Gregersen, E.M., Waseda, T., Parunov, J., Yim, S., Hirdaris, S., Ma, N., Guedes Soares, C., 2022. Uncertainties in long-term wave modelling. Marine Structures 84, 103217.

Buckley, M.P., Veron, F., 2016. Structure of the airflow above surface waves. Journal of Physical Oceanography 46, 1377–1397.

Boccotti, P., 2014. Wave mechanics and wave loads on marine structures. Elsevier.

Bonnefoy, F., Ducrozet, G., Le Touzé, D., Ferrant, P., 2010. Time domain simulation of nonlinear water waves using spectral methods. Advances in Numerical Simulation of Nonlinear Water Waves (World Scientific) 11, 129–164.

Chabchoub, A., 2016. Tracking Breather Dynamics in Irregular Sea State Conditions. Physical Review Letter 117, 144103.

Chabchoub, A., Hoffmann, N., Akhmediev, N., 2011. Rogue wave observation in a water wave tank. Physical Review Letters 106, 204502.

Chabchoub, A., Hoffmann, N., Akhmediev, N., 2012a. Observation of rogue wave holes in a water wave tank. Journal of Geophysical Research 117, C00J02.

Chabchoub, A., Akhmediev, N., Hoffmann, N., 2012b. Experimental study of spatiotemporally localized surface gravity water waves. Physical Review E 86: 016311.

Chabchoub, A., Waseda, T., Kibler, B., Akhmediev, N., 2017. Experiments on higher-order and degenerate Akhmediev breather-type rogue water waves. Journal of Ocean Engineering and Marine Energy 3, 385–394.

Chalikov, D., 2007. Numerical simulation of the Benjamin–Feir instability and its consequences. Physics of Fluids 19, 016602.

Chalikov, D., Sheinin, D., 1998. Direct modeling of one-dimensional nonlinear potential waves. Nonlinear Ocean Waves, edited by: Perrie, W., Advances in Fluid Mechanics 17, 207–258.

Chalikov, D., Sheinin, D., 2005. Modeling of extreme waves based on equations of potential flow with a free surface. Journal of Computational Physics 210, 247–273.

Cherneva, Z., Guedes Soares, C., 2008. Non-linearity and non-stationarity of the New Year abnormal wave. Applied Ocean Research 30, 215–220.

Cherneva, Z., Guedes Soares, C., 2011. Evolution of wave properties during propagation in a ship towing tank and an offshore basin. Ocean Engineering 38, 2254–2261.

Cherneva, Z., Tayfun, M.A., Guedes Soares, C., 2009. Statistics of nonlinear waves generated in an offshore wave basin. Journal of Geophysical Research, 114, C08005.

Cherneva, Z., Tayfun, M.A., Guedes Soares, C., 2013. Statistics of waves with different steepness simulated in a wave basin. Ocean Engineering, 60, 186–192.

Clamond, D., Grue, J., 2001. A fast method for fully nonlinear water wave computations. Journal of Fluid Mechanics 447, 337–355.

Clamond, D., Francius, M., Grue, J., Kharif, C., 2006. Long time interaction of envelope solitons and freak wave formations. European Journal of Mechanics B/Fluids 25, 536–553.

Clauss, G., 2002. Dramas of the sea: episodic waves and their impact on offshore structures. Applied Ocean Research 24, 147–161.

Clauss, G., Kosleck, S., Sprenger, F., Boeck, F., 2009. Adaptive Stretching of dynamic pressure distribution in Long- and short-crested sea states, in: Proceedings of the ASME 2009 28th International Conference on Ocean, Offshore and Arctic Engineering. Honolulu, OMAE2009–79485.

Clauss, G., Klein, M., Onorato, M., 2011. Formation of extraordinarily high waves in space and time. In: Proceedings of the 30th International Conference on Ocean, Offshore and Arctic Engineering, Rotterdam, The Netherlands, OMAE 2011–49545.

Constantin, A., 2016. Extrema of the dynamic pressure in an irrotational regular wave train. Physics of Fluids 28, 113604.

Constantin, A., Escher, J., Hsu, H.C., 2011. Pressure Beneath a Solitary Water Wave: Mathematical Theory and Experiments. Archive for Rational Mechanics and Analysis 201, 251–269.

Constantin, A., Strauss, W., 2010. Pressure beneath a stokes wave. Communications on Pure and Applied Mathematics 63, 533–557.

Cousins, W., Onorato, M., Chabchoub, A., Sapsis, T.P., 2019. Predicting ocean rogue waves from point measurements: An experimental study for unidirectional waves. Physical Review E 99, 032201.

Dias, F., Kharif, C., 1999. Nonlinear gravity and capillary gravity waves. Annual Review of Fluid Mechanics 31, 301–346.

Dommermuth, D.G., 2000. The initialization of nonlinear waves using an adjustment scheme. Wave Motion 32, 307–317.

Dommermuth, D.G., Yue, D.K., 1987. A high-order spectral method for the study of nonlinear gravity waves. Journal of Fluid Mechanics 184, 267–288.

Ducrozet, G., Bonnefoy, F., Le Touzé, D., Ferrant, P., 2016a. HOS-ocean: Open-source solver for nonlinear waves in open ocean based on High-Order Spectral method. Computer Physics Communication 203, 245–254.

Ducrozet, G., Bonnefoy, F., Ferrant, P., 2016b. On the equivalence of unidirectional rogue waves detected in periodic simulations and reproduced in numerical wave tanks. Ocean Engineering 117, 346–358.

Ducrozet, G., Bonnefoy, F., Le Touzé, D., Ferrant, P., 2012. A modified High-Order Spectral method for wavemaker modeling in a numerical wave tank. European Journal of Mechanics B/Fluid 34, 19–34.

Ducrozet, G., Bonnefoy, F., Touzé, D.L., Ferrant, P., 2007. 3-D HOS simulations of extreme waves in open seas. Natural Hazards and Earth System Sciences 7, 109–122.

Dysthe, K.B., 1979. Note on a modification to the nonlinear Schrödinger equation for application to deep water waves. Proceedings of the Royal Society of London A 369, 105–114.

Dysthe, K.B., Trulsen, K., Krogstad, H.E., Socquet-Juglard, H., 2003. Evolution of a narrow-band spectrum of random surface gravity waves. Journal of Fluid Mechanics 478, 1–10.

Dysthe, K.B., Krogstad, H.E, Müller, P., 2008. Oceanic rogue waves. Annual Review of Fluid Mechanics 40, 287–310.

Escher, J., Schlurmann, T., 2008. On the recovery of the free surface from the pressure within periodic traveling water waves. Journal of Nonlinear Mathematical Physics 15, 50–57.

Farazmand, M., Sapsis, T.P., 2017. Reduced-order prediction of rogue waves in two-dimensional deep-water waves. Journal of Computational Physics 340, 418–434.

Fedele, F., Tayfun, M.A., 2009. On nonlinear wave groups and crest statistics. Journal of Fluid Mechanics 620, 221–239.

Fedele, F., Cherneva, Z., Tayfun, M.A., Guedes Soares, C., 2010. Nonlinear Schrödinger invariants and wave statistics. Physics of Fluids 22, 036601.

Fedele, F., Lugni, C., Chawla, A., 2017. The sinking of the El Faro: Predicting real world rogue waves during Hurricane Joaquin. Scientific Report 7, 11188.

Forristall, G.Z., 2000. Wave crest distributions: observations and second-order theory. Journal of Physical Oceanography 30, 1931–1943.

Fructus, D., Clamond, D., Grue, J., Kristiansen, O., 2005. An efficient model for three-dimensional surface wave simulations. Part I: Free space problems. Journal of Computational Physics 205, 665–685.

Fujimoto, W., Waseda, T., Webb, A., 2018. Impact of the four-wave quasi-resonance on freak wave shapes in the ocean. Ocean Dynamics 69, 101–121.

Goda, Y., 2000. Random seas and design of maritime structures. Advanced Series on Ocean Engineering 15, World Scientific.

Goullet, A., Choi, W., 2011. A numerical and experimental study on the nonlinear evolution of long-crested irregular waves. Physics of Fluids 23, 016601.

Gramstad, O., Trulsen, K., 2007. Influence of crest and group length on the occurrence of freak waves. Journal of Fluid Mechanics 582, 463–472.

Guedes Soares, C., Cherneva, Z., Antão, E., 2003. Characteristics of abnormal waves in North Sea storm sea states. Applied Ocean Research 25 (6), 337–344.

Guedes Soares, C., Cherneva, Z., Antão, E., 2004. Abnormal waves during the hurricane Camille. Journal of Geophysical Research 109, C08008.

Guedes Soares, C., Cherneva, Z., Petrova, P., Antão, E., 2011. Large waves in sea states. In: Marine Technology and Engineering, Guedes Soares et al. (Eds), Taylor & Francis Group, London, 79–95.

Hagen, Ø., 2002. Statistics for the Draupner January 1995 freak wave event. In: Proceedings of the 21st International Conference on Offshore Mechanics and Arctic Engineering (OMAE'03), Oslo, Norway, OMAE2002–28608.

Holthuijsen, L.H., 2007. Waves in oceanic and coastal waters, Waves in Oceanic and Coastal Waters. Cambridge University Press. 10.1017/ CBO9780511618536.

Hu, Z., Zhang, X., Li, Y., Li, X., Qin, H., 2021. Numerical simulations of super rogue waves in a numerical wave tank. Ocean Engineering 229, 108929.

Iafrati, A., Babanin, A., Onorato, M., 2013. Modulational instability, wave breaking, and formation of large-scale dipoles in the atmosphere. Physical Review Letter 110, 184504.

Janssen, P.A.E.M., 2003. Nonlinear four-wave interactions and freak waves. Journal of Physical Oceanography 33, 863–884.

Janssen, P.A.E.M., Bidlot, J., 2009. On an extension of the freak wave warning system and its verification, ECMWF Tech. Memo. 588, Europe Centre for Medium-Range Weather Forecasts, Reading, UK.

Kit, E., Shemer, L., 2002. Spatial versions of the Zakharov and Dysthe evolution equations for deep water gravity waves. Journal of Fluid Mechanics 450, 201–205.

Kharif, C., Pelinovsky, E., 2003. Physical mechanisms of the rogue wave phenomenon. European Journal of Mechanics - B/Fluids 21, 561–577.

Kharif, C., Pelinovsky, E., Slunyaev, A., 2009. Rogue waves in the ocean. Springer–Verlag Berlin Heidelberg, Germany.

Klein, M., Clauss, G.F., Rajendran, S., Guedes Soares, C., Onorato, M., 2016. Peregrine breathers as design waves for wave-structure interaction. Ocean Engineering 128, 199–212.

Kogelbauer, F., 2015. Recovery of the wave profile for irrotational periodic water waves from pressure measurements. Nonlinear Analysis Real World Applications 22, 219–224.

Kuznetsov, E., 1977. Solitons in a parametrically unstable plasma. In: Akademiia Nauk SSSR Doklady 139, Springer.

Lawton, G., 2001. Monsters of the deep (The Perfect Wave). New Scientist 170, 28–32.

Lo, E., Mei, C.C., 1985. A numerical study of water-wave modulation based on a higher-order nonlinear Schrödinger equation. Journal of Fluid Mechanics 150, 395–416.

Longuet-Higgins, M.S., 1983. On the joint distribution of wave periods and amplitudes in a random wave field. Proceedings of the Royal Society, London, A389, 241–258.

Ma, Y., 1979. The perturbed plane-wave solutions of the cubic Schrodinger equation. Studies in Applied Mathematics 60, 43–58.

Mei, C.C., 1989. The Applied Dynamics of Ocean Surface Waves. World Scientific, Singapore.

Mori, N., Janssen, P.A.E.M., 2006. On kurtosis and occurrence probability of freak waves. Journal of Physical Oceanography 36, 1471–1483.

Mori, N., Onorato, M., Janssen, P.A., Osborne, A.R., Serio, M., 2007. On the extreme statistics of long-crested deep water waves: Theory and experiments. Journal of Geophysical Research 112, C09011.

Mori, N., Onorato, M., Janssen, P.A., 2011. On the estimation of the kurtosis in directional sea states for freak wave forecasting. Journal of Physical Oceanography 41, 1484–1497.

Olagnon, M., 2000. Vagues extrêmes – Vagues scélérates. http://www.ifremer.fr/web-com/molagnon/jpo2000/.

Olfateh, M., Ware, P., Callaghan, D.P., Nielsen, P., Baldock, T.E., 2017. Momentum transfer under laboratory wind waves. Coastal Engineering 121, 255–264.

Oliveras, K.L., Vasan, V., Deconinck, B., Henderson, D., 2012. Recovering the water-wave profile from pressure measurements. SIAM Journal on Applied Mathematics 72, 897–918.

Onorato, M., Suret, P., 2016. Twenty years of progresses in oceanic rogue waves: the role played by weakly nonlinear models. Natural Hazards 84, 541–548.

Onorato, M., Osborne, A., Serio, M., Bertone, S., 2001. Freak waves in random oceanic sea states. Physical Review Letters 86(25), 5831–5834.

Onorato, M., Osborne, A., Serio, M., Cavaleri, L., 2005. Modulational instability and non-Gaussian statistics in experimental random water-wave trains. Physics of Fluids 17, 078101.

Onorato M., Osborne A., Serio M., Cavaleri L., Brandini C., Stansberg C., 2006. Extreme waves, modulational instability and second order theory: Wave flume experiments on irregular waves, European Journal of Mechanics B, 25, 583–601.

Onorato, M., Cavaleri, L., Fouques, S, Gramstad, O., Janssen, P.A.E.M., Monbaliu, J., Osborne, A., Pakozdi, C., Serio, M., Stansberg, C., Toffoli, A., Trulsen, K., 2009. Statistical properties of mechanically generated surface gravity waves: a laboratory experiment in three-dimensional wave basin. Journal of Fluid Mechanics 627, 235–257.

Onorato, M., Residori, S., Bortolozzo, U., Montina, A., Arecchi, F., 2013a Rogue waves and their generating mechanisms in different physical contexts. Physics Reports 528, 47–89.

Onorato, M., Proment, D., Clauss, G., Klein, M., 2013b. Rogue Waves: from nonlinear Schrödinger breather solutions to sea-keeping test. PLoS ONE 8(2), e54629.

Onorato, M., Proment, D., Toffoli, A., 2011. Triggering rogue waves in opposing currents. Physical Review Letters 107, 184502.

Onorato, M., Cavaleri, L., Randoux, S., Suret, P., Ruiz, M.I., Alfonso, M. De, Benetazzo, A., 2021. Observation of a giant nonlinear wave - packet on the surface of the ocean. Scientific Reports 11, 23606.

Osborne, A.R., 2010. Nonlinear Ocean Wave and the Inverse Scattering Transform. Elsevier, New York.

Peregrine, D.H., 1983. Water waves, nonlinear Schrödinger equations and their solutions. The Journal of Australian Mathematical Society Series B 25, 16–43.

Perić, R., Hoffmann, N., Chabchoub, A., 2015. Initial wave breaking dynamics of Peregrine-type rogue waves: a numerical and experimental study. European Journal of Mechanics-B/Fluids 49, 71–76.

Petrova, P., Cherneva, Z., Guedes Soares, C., 2007. On the adequacy of second-order models to predict abnormal waves. Ocean Engineering 34, 956–961.

Petrova, P.G., Arena, F., Guedes Soares, C., 2011. Space-time evolution of random wave groups with high waves based on the quasi-determinism theory. Ocean Engineering 38, 1640–1648.

Ponce de Leon, S., Guedes Soares, C., 2022. Distribution of average extreme wave parameters in the North Atlantic from numerical simulations. Ocean Engineering 253, 110901.

Rodriguez, G.R., Guedes Soares, C., 2001. Correlation between successive wave heights and periods in mixed sea states. Ocean Engineering 28(8), 1009–1030.

Seiffert, B.R., Ducrozet, G., 2018. Simulation of breaking waves using the high-order spectral method with laboratory experiments: wave-breaking energy dissipation. Ocean Dynamics 68, 65–89.

Seiffert, B.R., Ducrozet, G., Bonnefoy, F., 2017. Simulation of breaking waves using the high-order spectral method with laboratory experiments: Wave-breaking onset. Ocean Modelling 119, 94–104.

Shemer, L., Alperovich, L., 2013. Peregrine breather revisited. Physics of Fluids 25, 051701.

Shemer, L., Dorfman, B., 2008. Experimental and numerical study of spatial and temporal evolution of nonlinear wave groups. Nonlinear Processes in Geophysics 15, 931–942.

Shemer, L., Ee, B.K., 2015. Steep unidirectional wave groups - fully nonlinear simulations vs. Experiments. Nonlinear Processes in Geophysics Discussions 22, 737–747.

Shemer, L., Sergeeva, A., 2009. An experimental study of spatial evolution of statistical parameters in a unidirectional narrow-banded random wave field, Journal of Geophysical Research 114, C01015.

Shemer, L., Kit, E., Jiao, H., 2002. An experimental and numerical study of the spatial evolution of unidirectional nonlinear water-wave groups. Physics of Fluids 14, 3380–3390.

Shemer, L., Sergeeva, A., Slunyaev, A., 2010a. Applicability of envelope model equations for simulation of narrow-spectrum unidirectional random wave field evolution: Experimental validation. Physics of Fluids 22, 016601.

Shemer, L., Sergeeva, A., Liberzon, D., 2010b. Effect of the initial spectrum on the spatial evolution of statistics of unidirectional nonlinear random waves. Journal of Geophysical Research 115, C12039.

Slunyaev, A., 2018. Group-wave resonances in nonlinear dispersive media: The case of gravity water waves. Physical Review E 97, 010202.

Slunyaev, A., Dosaev, A., 2019. On the incomplete recurrence of modulationally unstable deep-water surface gravity waves. Communications in Nonlinear Science and Numerical Simulation 66, 167–182.

Slunyaev, A., Shrira, V.I., 2013. On the highest non-breaking wave in a group: Fully nonlinear water wave breathers versus weakly nonlinear theory. Journal of Fluid Mechanics 735, 203–248.

Slunyaev, A., Didenkulova, I., Pelinovsky, E., 2011. Rogue waters. Contemporary Physics 52, 571–590.

Slunyaev, A., Pelinovsky, E., Guedes Soares, C., 2005. Modeling Freak Waves from the North Sea. Applied Ocean Research 27, 12–22.

Slunyaev, A., Pelinovsky, E., Soares, C.G., 2014. Reconstruction of extreme events through numerical simulations. Journal of Offshore Mechanics and Arctic Engineering 136(1), 011302.

Slunyaev, A., Sergeeva, A., Didenkulova, I., 2016. Rogue events in spatiotemporal numerical simulations of unidirectional waves in basins of different depth. Natural Hazards 84, 549–565.

Socquet-Juglard, H., Dysthe, K., Trulsen, K., Krogstad, H.E., Liu, J., 2005. Distribution of surface gravity waves during spectral changes. Journal of Fluid Mechanics 542, 195–216.

Stansberg C.T., 1994. Effects from directionality and spectral bandwidth on non-linear spatial modulations of deep-water surface gravity waves. In Proceedings of the 24th International Conference on Coastal Engineering, Kobe, Japan, 579–593.

Stansell, P., Wolfram, J., Linfoot, B., 2004. Improved joint probability distribution for ocean wave heights and periods. Journal of Fluid Mechanics 503, 273–297.

Stiassnie, M., Shemer, L., 1987. Energy computations for evolution of class I and class II instabilities of Stokes waves. Journal of Fluid Mechanics 174, 299–312.

Su, M.Y., 1982. Evolution of groups of gravity waves with moderate to high steepness. Physics of Fluids 25, 2167–2174.

Tanaka, M., 1990. Maximum amplitude of modulated wave train. Wave Motion 12, 559–568.

Tanaka, M., 2001a. A method of studying nonlinear random field of surface gravity waves by direct numerical simulation. Fluid Dynamics Research 28, 41–60.

Tanaka, M., 2001b. Verification of Hasselmann's energy transfer among surface gravity waves by direct numerical simulations of primitive equations. Journal of Fluid Mechanics 444, 199–221.

Tayfun, M.A., 2006. Statistics of nonlinear wave crests and groups. Ocean Engineering 33, 1589–1622.

Tayfun, M.A., Fedele, F., 2007. Wave-height distributions and nonlinear effects. Ocean Engineering 34, 1631–1649.

Taylor, P.H., Adcock, T., Borthwick, A., Walker, D., Yao, Y., 2006. The nature of the Draupner giant wave of 1st January 1995 and associated sea-state, and how to estimate directional spreading from an Eulerian surface elevation time history. In: the 9th International Workshop on Wave Hindcasting and Forecasting, Victoria, Canada.

Toffoli, A., Lefevre, J.M., Bitner-Gregersen, E., Monbailiu, J., 2005. Towards the identification of warning criteria: Analysis of a ship accident database. Applied Ocean Research 27, 281–291.

Toffoli, A., Bitner-Gregersen, E.M., Osborne, A.R., Serio, M., Monbaliu, J., Onorato, M., 2011. Extreme waves in random crossing seas: Laboratory experiments and numerical simulations. Geophysical Research Letters 38, L06605.

Toffoli, A., Bitner-Gregersen, E.M., Onorato, M., Babanin, A.V., 2008a. Wave crest and trough distributions in a broad-banded directional wave field. Ocean Engineering 35, 1784–1792.

Toffoli, A., Onorato, M., Bitner-Gregersen, E.M., Osborne, A.R., Babanin, A.V., 2008b. Surface gravity waves from direct numerical simulations of the Euler equations: A comparison with second-order theory. Ocean Engineering 35, 367–379.

Toffoli, A., Gramstad, O., Trulsen, K., Monbaliu, J., Bitner-Gregersen, E.M., Onorato, M., 2010a. Evolution of weakly nonlinear random directional waves: laboratory experiments and numerical simulations. Journal of Fluid Mechanics 664, 313–336.

Toffoli, A., Babanin, A.V., Onorato, M., Waseda, T., 2010b. Maximum steepness of oceanic waves: Field and laboratory experiments, Geophysical Research Letter, 37, L05603.

Toffoli, A., Waseda, T., Houtani, H., Cavaleri, L., Greaves, D., Onorato, M., 2015. Rogue waves in opposing currents: An experimental study on deterministic and stochastic wave trains. Journal of Fluid Mechanics 769, 277–297.

Toffoli, A., Proment, D., Salman, H., Monbaliu, J., Frascoli, F., Dafilis, M., Stramignoni, E., Forza, R., Manfrin, M., Onorato, M., 2017. Wind generated rogue waves in an annular wave flume. Physical Review Letters 118, 144503.

Touboul, J., Pelinovsky, E., 2018. On the use of linear theory to estimate bottom pressure distribution under nonlinear surface waves. European Journal of Mechanics - B/Fluids 67, 97–103.

Trulsen, K., Dysthe, K.B., 1996. A modified nonlinear Schrödinger equation for broader bandwidth gravity waves on deep water. Wave Motion 24, 281–289.

Trulsen, K., Dysthe, K.B., 1997. Frequency downshift in three-dimensional wave trains in a deep basin. Journal of Fluid Mechanics 352, 359–373.

Trulsen, K., Kliakhandler, I., Dysthe, K.B., Velarde, M.G., 2000. On weakly nonlinear modulation of waves on deep water. Physics of Fluids 12, 2432–2437.

Tsai, C.H., Huang, M.C., Young, F.J., Lin, Y.C., Li, H.W., 2005. On the recovery of surface wave by pressure transfer function. Ocean Engineering 32, 1247–1259.

Veltcheva, A., Guedes Soares, C., 2016. Nonlinearity of abnormal waves by the Hilbert-Huang Transform method. Ocean Engineering 115, 30–38.

Waseda, T., 2005. Experimental investigation and applications of the modulational wave train. In: Proceedings of the Workshop on Rogue Waves, 12–15 December 2005, ICMS, Edinburgh.

Waseda, T., Fujimoto, W., Chabchoub, A., 2019. On the asymmetric spectral broadening of a hydrodynamic modulated wave train in the optical regime. Fluids 4(2), 84.

Waseda, T., Kinoshita, T. and Tamura, H., 2009. Evolution of a Random Directional Wave and Freak Wave Occurrence. Journal of Physical Oceanography 39, 621–639.

West, B.J., Brueckner, K.A., Jand, R.S., Milder, D.M., Milton, R.L., 1987. A new numerical method for surface hydrodynamics. Journal of Geophysical Research 92 (C11), 11803–11824.

Xiao, W.T., Liu, Y.M., Wu, G.Y., Yue, D.K.P., 2013. Rogue wave occurrence and dynamics by direct simulations of nonlinear wave-field evolution. Journal of Fluid Mechanics 720, 357–392.

Yildirim, B., Karniadakis, G.E., 2015. Stochastic simulations of ocean waves: An uncertainty quantification study. Ocean Model 86, 15–35.

Yuen, H.C., Lake, B.M., 1980. Instabilities of waves on deep water. Annual Review of Fluid Mechanics 12, 303–334.

Zakharov, V.E., Ostrovsky, L.A., 2009. Modulation instability: The beginning. Physica D 238, 540–548.

Zhang, H.D., Cherneva, Z., Guedes Soares, C., 2013. Joint distributions of wave height and period in laboratory generated nonlinear sea states. Ocean Engineering 74, 72–80.

Zhang, H.D., Guedes Soares, C., Onorato, M., 2014. Modelling of the spatial evolution of extreme laboratory wave heights with the nonlinear Schrödinger and Dysthe equations. Ocean Engineering 89, 1–9.

Zhang, H.D., Guedes Soares, C., Onorato, M., 2015. Modelling of the spatial evolution of extreme laboratory wave crest and trough heights with the NLS-type equations. Applied Ocean Research 52, 140–150.

Zhang, H.D., Guedes Soares, C., Chalikov, D., Toffoli, A., 2016a. Modelling the spatial evolutions of nonlinear unidirectional surface gravity waves with fully nonlinear numerical method. Ocean Engineering 125, 60–69.

Zhang, H.D., Guedes Soares, C., Onorato, M., Toffoli, A., 2016b. Modelling of the temporal and spatial evolutions of weakly nonlinear random directional waves with the modified nonlinear Schrödinger equations. Applied Ocean Research 55, 130–140.

Zhang, H.D., Ducrozet, G., Klein, M., Guedes Soares, C., 2017a. An experimental and numerical study on Breather solutions for surface waves in the intermediate water depth. Ocean Engineering 133, 262–270.

Zhang, H.D., Sanina, E., Babanin, A., Guedes Soares, C., 2017b. On the analysis of 2D nonlinear gravity waves with a fully nonlinear numerical model. Wave Motion 70, 152–165.

Zhang, H.D., Shi, H.D., Guedes Soares, C., 2019. Evolutionary properties of mechanically generated deepwater extreme waves induced by nonlinear wave focusing. Ocean Engineering 186, 106077.

Zhang, H.D., Wang, X.J., Shi, H.D., Guedes Soares, C., 2021. Investigation on abnormal wave dynamics in regular and irregular sea states. Ocean Engineering 222, 108602.

Zhang, H.D., Liao, X.M., Shi, H.D., Babanin, A., Guedes Soares, C., 2022. Effect of initial condition uncertainty on the profile of maximum wave. Marine Structures 82, 103127.

Zhang, H.D., Liao, X.M., Xin, Z.X., Shi, H.D., Guedes Soares, C., 2023. Experimental study on dynamic pressure under the crest of unidirectional nonlinear waves. Ocean Engineering 276, 114251.

MIX
Papier aus verantwortungsvollen Quellen
Paper from responsible sources
FSC® C105338

If you have any concerns about our products,
you can contact us on
ProductSafety@springernature.com

In case Publisher is established outside the EU,
the EU authorized representative is:
Springer Nature Customer Service Center GmbH
Europaplatz 3, 69115 Heidelberg, Germany

Printed by Libri Plureos GmbH
in Hamburg, Germany